Thomas Villmann · Marika Kaden ·
Tina Geweniger · Frank-Michael Schleif
Editors

Advances in Self-Organizing Maps, Learning Vector Quantization, Interpretable Machine Learning, and Beyond

Proceedings of the 15th International Workshop, WSOM+ 2024, Mittweida, Germany, July 10–12, 2024

Springer

Editors
Thomas Villmann
Hochschule Mittweida
Mittweida, Germany

Tina Geweniger
Westsächsische Hochschule Zwickau
Zwickau, Germany

Marika Kaden
Hochschule Mittweida
Mittweida, Germany

Frank-Michael Schleif
Technische Hochschule Würzburg-Schweinf
Würzburg, Germany

ISSN 2367-3370 ISSN 2367-3389 (electronic)
Lecture Notes in Networks and Systems
ISBN 978-3-031-67158-6 ISBN 978-3-031-67159-3 (eBook)
https://doi.org/10.1007/978-3-031-67159-3

This Springer imprint is published by the registered company Springer Nature Switzerland AG
The registered company address is: Gewerbestrasse 11, 6330 Cham, Switzerland

Lecture Notes in Networks and Systems 1087

The series "Lecture Notes in Networks and Systems" publishes the latest developments in Networks and Systems—quickly, informally and with high quality. Original research reported in proceedings and post-proceedings represents the core of LNNS.

Volumes published in LNNS embrace all aspects and subfields of, as well as new challenges in, Networks and Systems.

The series contains proceedings and edited volumes in systems and networks, spanning the areas of Cyber-Physical Systems, Autonomous Systems, Sensor Networks, Control Systems, Energy Systems, Automotive Systems, Biological Systems, Vehicular Networking and Connected Vehicles, Aerospace Systems, Automation, Manufacturing, Smart Grids, Nonlinear Systems, Power Systems, Robotics, Social Systems, Economic Systems and other. Of particular value to both the contributors and the readership are the short publication timeframe and the world-wide distribution and exposure which enable both a wide and rapid dissemination of research output.

The series covers the theory, applications, and perspectives on the state of the art and future developments relevant to systems and networks, decision making, control, complex processes and related areas, as embedded in the fields of interdisciplinary and applied sciences, engineering, computer science, physics, economics, social, and life sciences, as well as the paradigms and methodologies behind them.

Indexed by SCOPUS, INSPEC, WTI Frankfurt eG, zbMATH, SCImago.

All books published in the series are submitted for consideration in Web of Science.

For proposals from Asia please contact Aninda Bose (aninda.bose@springer.com).

Preface

This book contains the peer-reviewed contributions presented at the 15th Workshop on Self-Organizing Maps, Learning Vector Quantization and Beyond (WSOM+ 2024), held at the University of Applied Sciences Mittweida (UAS Mittweida), Germany, on July 10–12, 2024. It was mainly organized by the Saxon Institute for Computational Intelligence and Machine Learning (SICIM) of the UAS Mittweida.

This conference series began in 1995 in Espoo, Finland, initiated by Teuvo Kohonen, the inventor of the Self-Organizing Map (SOM) and Learning Vector Quantization (LVQ) models for high-dimensional data analysis, visualization, and classification. Since then, the biennial international conference has traveled around the world with stops in Paris, Santiago de Chile, Houston, and Prague, to name a few. This year, the 15th WSOM returned to Mittweida after the 10th workshop in 2014.

More than thirty-five years after the first publications on SOM and LVQ, the attraction of these approaches remains undiminished and new variants and extensions are popular. In particular, the obvious interpretability of these prototype-based methods together with their robustness and computational sparseness provide many opportunities for advanced machine learning tasks in a wide variety of applications ranging from biomedicine, space science, engineering to economics and social sciences, for example. The flexibility and simplicity of LVQ and SOM also allows the integration of modern aspects such as deep architectures, probabilistic methods, and reasoning as well as relevance learning.

The WSOM+ 2024 conference brings together researchers and practitioners in the field of self-organizing systems, interpretable and robust machine learning models related to SOM and LVQ and beyond. In these proceedings, the reader will find selected papers from the conference. These articles reflect new theoretical developments, advanced applications, and new methodological insights.

Thus, this book is recommended for researchers and practitioners in data analytics and machine learning, especially those who are interested in the latest developments in interpretable and robust unsupervised learning, data visualization, classification, and self-organization.

We warmly thank the members of the Steering Committee and the Executive Committee. Our sincere thanks go to Barbara Hammer, John Aldo Lee, and Peter Tino for their plenary talks. We are grateful to the members of the Program Committee and other reviewers for their excellent and timely work, and above all to the authors whose contributions made this book possible.

July 2024

Thomas Villmann
Frank-Michael Schleif
Marika Kaden
Tina Geweniger

Organization

WSOM+ 2024 as the fifteenth of the WSOM conference series was held during July 10–12, 2024 in Mittweida, Saxony (Germany). It was organized by the Saxon Institut of Computational Intelligence and Machine Leraning (SICIM) of the University of Applied Sciences Mittweida in cooperation with the Faculty for Applied Computer and Biosciences and the Mittweida Institute for Computational Intelligence and Intelligent Data Analysis e.V. .

Executive Committee

General Chair

Thomas Villmann	SICIM, University of Applied Sciences Mittweida, Germany

Program Chair

Frank-M. Schleif	Technical University of Würzburg-Schweinfurt, Germany

Local Chairs

Marika Kaden	SICIM, University of Applied Sciences Mittweida, Germany
Tina Geweniger	University of Applied Sciences Zwickau, Germany

Local Organizer

Matthias Baumgart und Annett Kober	University of Applied Sciences Mittweida, Germany

Program Committee

Backhaus, Andreas	Compolytics GmbH Barleben, Germany
Barreto, Guilherme A.	Federal University of Ceará, Brasil

Biehl, Michael	University Groningen, The Netherlands
Coelho, David Nascimento	Federal University of Ceará, Brasil
Corona, Francesco	Aalto University, Finland
de Vries, Gert-Jan	Philips, The Netherlands
Engelsberger, Alexander	University of Applied Sciences Mittweida, Germany
Gepperth, Alexander R.T.	University of Applied Sciences Fulda, Germany
Geweniger, Tina	University of Applied Sciences Zwickau, Germany
Girau, Bernard	University of Lorraine, France
Hartono, Pitoyo	Chukyo University, Japan
Kaden, Marika	University of Applied Sciences Mittweida, Germany
Lebbah, Mustapha	Université Paris Saclay, France
Lee, John A.	Chatholic University Louvain, Belgium
Lisboa, Paulo	Liverpool John Moores University, UK
Miikkulainen, Risto	The University of Texas at Austin, USA
Mulders, Dounia	Chatholic University Louvain, Belgium
Nebel, David	University of Applied Sciences Mittweida, Germany
Neme, Antonio	Universidad Nacional Autónoma de México, Mexico
Olteanu, Madalina	University Paris Dauphine PSL, France
Ortega-Martorell, Sandra	Liverpool John Moores University, UK
Resta, Marina	University of Genova, Italy
Saralajew, Sascha	NEC Laboratories Europe GmbH Heidelberg, Germany
Schleif, Frank-Michael	Technical University of Würzburg-Schweinfurt, Germany
Schulz, Alexander	Bielefeld University, Germany
Seiffert, Udo	Compolytics GmbH Barleben, Germany
Silva, Leandro Augusto	Mackenzie Presbyterian University, Brazil
Taylor, Joshua Jordan	The University of Texas at Austin, USA
Van Hulle, Marc	Chatholic University Louvain, Belgium
Vaquet, Valerie	Bielefeld University, Germany
Vellido, Alfredo	Universitat Politècnica de Catalunya, Espania
Villmann, Thomas	University of Applied Sciences Mittweida, Germany

Steering Committee

Marie Cottrell	Université Paris 1, Pantheón-Sorbonne, France
France Pablo Estévez	University of Chile, Chile
Jan Faigl	Czech Technical University in Prague, Czech
Sami Kaski	Aalto University, Finland
Jean-Charles Lamirel	Loria, France
Thomas Martinetz	University Lübeck, Germany
Erzsébet Merényi	Rice University Houston, USA
Erkki Oja	Aalto University, Finland
Madalina Olteanu	University Paris Dauphine PSL, France
José Príncipe	University of Florida, USA
Helge Ritter	Bielefeld University, Germany
Takaeshi Yamakawa	Kyushu Institute of Technology, Japan
Hujun Yin	University of Manchester, UK
Michel Verleysen	Université Catholique de Louvain, Belgium
Alfredo Vellido	Polytechnic University of Catalonia, Spain
Thomas Villmann	University of Applied Sciences Mittweida, Germany

In Memorian

Teuvo Kohonen	Aalto University Espoo, Finland

Sponsors and Supporters

We are grateful for the support for WSOM+ 2024 and would like to thank the following sponsors:

- Basycon Unternehmensberatung GmbH, Munich, Germany
- Mittweida Institute for Computational Intelligence and Intelligent Data Analysis e.V. (CIID)

Further, we thank Alexander Engelsberger for the design of the WSOM+ 2024 conference logo.

Contents

New Cloth Unto an Old Garment: SOM for Regeneration Learning

Rewbenio A. Frota[1,3(✉)], Guilherme A. Barreto[2], Marley M. B. R. Vellasco[3], and Candida Menezes de Jesus[1]

[1] Petróleo Brasileiro S.A. - PETROBRAS, Rio de Janeiro, RJ, Brazil
rewbenio@petrobras.com.br
[2] Graduate Program in Teleinformatics Engineering, Center of Technology, Federal University of Ceará (UFC), Fortaleza, CE, Brazil
[3] Department of Electrical Engineering, Pontifical Catholic University of Rio de Janeiro (PUC-Rio), Rio de Janeiro, RJ, Brazil

Abstract. A recent paradigm called *Regeneration Learning* addresses generative problems where the target data (e.g., images) is more complex than the available input source. While current cross-modal representation and regeneration learning rely on supervised deep learning models, this paper aims to revisit the adequacy of unsupervised models in this field. In this regard, we propose a new unsupervised approach that utilizes the SOM as a heteroassociative memory model to learn cross-modal representations in a topologically coherent map. This approach enables bidirectional predictive/regenerative mapping between domains. We evaluate the potential of this method for an unsolved (so far!) practical problem in petroleum geoscience.

Keywords: Self-Organizing Maps · Regeneration learning · Variational Auto-encoders · Petrophysical Logs · Acoustic Image Logs

1 Introduction

This paper proposes an unsupervised framework based on Self-Organizing Maps (SOM) [6] to allow data generation (or rather regeneration [9]) throughout heterogeneous data domains by mapping multimodal representations in a coherent topology-preserving structure. By a simple modification [8], which we revisit and expand here, the basic SOM can learn nonlinear input-output mappings with no constraints on the data types or imbalances in the information content of the data. The proposed modification allows the SOM to perform a heteroassociative mapping [6] and function as a cross-modal estimator.

Using SOM as an estimator has succeeded in control engineering and system identification [2], and in time-series forecasting [1]. In [7], the SOM was used to build a probabilistic heteroassociative memory to learn and classify multimodal patterns. Thus, we expand the scope of this mapping by leveraging the classic

T. Villmann et al. (Eds.): WSOM+2024, LNNS 1087, pp. 1–10, 2024.
https://doi.org/10.1007/978-3-031-67159-3_1

input-output relationship to a broader scope of cross-modal data modeling, such as spatial or semantic interdependencies in multimodal data sources.

The proposed framework allows the SOM to perform *source-to-representation* learning in the context of the recently formalised *Regeneration Learning* (RgL) paradigm [9]. RgL aims to deal with some of the challenges for conditional generative models, especially when the target data Y (e.g., images) has a much higher dimensionality than the correspondent input source X. RgL tackles this information mismatch problem employing a "divide to conquer" strategy summarized in three stages: (1) learn a compact representation Z from target data Y (target-to-representation stage); (2) Learn to predict Z from X (source-to-representation stage) and (3) Learn to regenerate Y from X, via Z (regeneration stage).

Usually, the source-to-representation stage relies on complex deep learning models. As an alternative, we hypothesize that a fully unsupervised framework could perform similarly to supervised-based models in the regeneration learning paradigm. Therefore, our proposal places unsupervised learning, instantiated here by the SOM algorithm, at the core of this relevant field. In the remaining stages, we complement our proposal by applying an unsupervised model, the β-VAE [4]. However, the framework is flexible enough to use other generative models. The main advantages of the proposed methodology are threefold:

- It is entirely unsupervised, specially in the source-to-representation stage, which often relies on computationally heavy deep learning models.
- Valid local relations encoded in the SOM prototypes free the proposed methodology from the burden of learning an explicit nonlinear regression model for the source-to-representation stage.
- After learning the source-to-representation mapping, the reverse mapping from Y to X is straightforward. No further learning is required.

We investigate the potential of the proposed framework in a yet unsolved complex cross-modal generative data problem in the petroleum geoscience speciality of petrophysics.

2 Proposed Framework

In this paper, we leverage a conceptual modification proposed in the 1980s enabling SOM to learn input-output mappings (Fig. 1) by means of incorporating input and corresponding output information in an augmented input vector $\mathbf{x}_{aug} = \left[\mathbf{x}^{in}, \mathbf{x}^{out}\right]$. Such an augmented input makes SOM to automatically encode the input-output mapping into its units' prototypes[1]: $\mathbf{w}_i = \left[\mathbf{w}_i^{in}, \mathbf{w}_i^{out}\right]$. That said, our proposal generalizes this concept by considering a source-to-target data heteroassociative mapping, leveraging SOM for cross-modal generative/regenerative learning.

$$\mathbf{x}_{aug} = \left[\mathbf{x}^s, \mathbf{x}^t\right] = [x_1^s, \ \ldots, \ x_q^s, \ x_1^t, \ \ldots, \ x_r^t], \tag{1}$$

[1] i refers to an unit in the SOM, while i^* refers to the winning unit.

q and r denote the dimensions of source (s) and target (t) data, respectively. Thus, the SOM's prototypes are expressed as $\mathbf{w}_i = [\mathbf{w}_i^s, \mathbf{w}_i^t]$. We use the term *Expanded Regenerative SOM* (ERGSOM) to refer to a map trained with augmented inputs, as defined in Eq. 1.

To apply the heteroassociative mapping proposed here, a source space χ and a target space $\varUpsilon$ are considered. They are respectively composed by different modality data, corresponding to related contexts in the real world. They may be two different data descriptions of the same object, sharing semantic information despite potential differences in complexity or recording formats. Good examples are an audio voice message and its description in a particular idiom; or, as we will later show in petrophysics, borehole image logs and other conventional logs acquired in the same well. In such scenarios, our aim is to apply the ERGSOM in heteroassociative source-to-target mapping, allowing multimodal cross-representation learning [13].

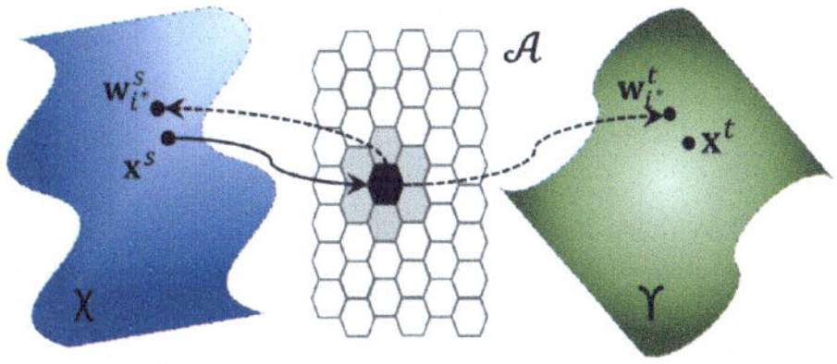

Fig. 1. Illustration of the ERGSOM-based source-to-target heteroassociative mapping.

Figure 1 illustrates the ERGSOM model for source-to-target heteroassociative mapping. The map $\mathcal{A}$ links different data spaces χ and $\varUpsilon$ simply by sharing the location of the winning unit i^*, which is used to retrieve $\mathbf{w}_{i^*}^t$ from $\mathbf{x}^s$. The framework based on ERGSOM can be summarized as follows[2]:

Stage 1: Learn a compact representation Z of the target space $\varUpsilon$ from Y. This stage will use a custom Convolutional β-VAE (β-CVAE)[3].
Stage 2: Learn the associative mapping of X and Z with ERGSOM.
Stage 3: Learn to regenerate[4] Y from X using the ERGSOM mapping. This stage is already granted by the use of the β-CVAE model built in Stage 1.

---

[2] We apply the following notation in this paper: (i) $\mathbf{x} \in \mathrm{I\!R}^q$, in the source space χ, so that $X = [\mathbf{x}_1 \ \dots \ \mathbf{x}_N]$; ($ii$) $\mathbf{y} \in \varUpsilon$, composing the target dataset is $Y = [\mathbf{y}_1 \ \dots \ \mathbf{y}_N]$; ($iii$) $\mathbf{z} \in \mathrm{I\!R}^r$, the target representation in space $\mathcal{Z}$, so that $Z = [\mathbf{z}_1 \ \dots \ \mathbf{z}_N]$ and (iv) N is the number of data vectors.

[3] Python 3.10.9 is used for experiments. We employ open source libraries such as Scikit-Learn, Keras, Tensorflow, Pandas and Numpy. The SOM is implemented using the IntraSOM package [3].

[4] Regenerative learning happens when ERGSOM feeds a generative model to regenerate complex data from simpler inputs.

Stage 4 (optional): Without any further changes, predict X from Y, as the heteroassociative mapping (Stage 2) allows bidirectional inference.

The fourth stage of our proposal shows that it goes beyond a standard regeneration learning procedure (Stages 1 to 3). The ERGSOM also embeds in the methodology a *data understanding* feature (in the terms stated in [9]).

3 Complex Data Regeneration with ERGSOM

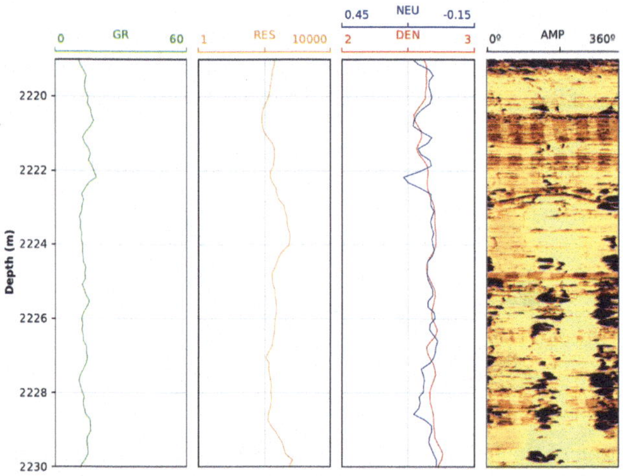

Fig. 2. Four basic logs (*GR*, *RES*, *DEN* and *NEU* in tracks 1 to 3) and an acoustic image log (*AMP* in track 4) in a portion of Well D.

Our application lies in the field of petroleum geoscience called petrophysics, which mainly deals with different measurements taken inside a just-drilled well, known as petrophysical *logs*. Logs can be either basic ones as density (*DEN*) and gamma ray (*GR*), or special ones as acoustic image logs. Special logs are usually much more expensive to acquire. More details about petrophysical logs and their physical working principle can be found in [10]. Figure 2 illustrates this kind of data, showing four conventional logs and an acoustic image log[5].

[5] Our dataset was acquired from five wells (named A to E) drilled in carbonate reservoirs in the Brazilian pre-salt. It consists of high resolution acoustic images with a corresponding set of basic open hole log curves: caliper of borehole (*CAL*), which is used only for data quality verification; gamma ray log *GR*, bulk density (*DEN*); neutron-porosity log (*NEU*); transit-time of compressional waves (*DTC*); photoelectric factor (*PE*); formation resistivity (*RES*) and total porosity measured by NMR (*nmrPhiT*). We apply the log transformation to the RES curve. The wells were split into five training/validation and test sets. Every training set consists of four wells, while the remaining one is kept for blind testing (e.g. results shown for Well A were obtained from models trained with data from Wells B to E). The *Wellbore Acoustic Image Database* (WAID) is available at https://github.com/petrobras/WAID.

It is evident from Fig. 2 that the image logs carry high-density information about the borehole wall. In contrast, conventional logs have a lower content of information. Obtaining valuable information from cheaper and simpler data sources, such as conventional logs, offers several advantages. It not only helps to reduce data acquisition costs but also enables the anticipation of critical information that can accelerate decision-making processes. We explore the potential of the ERGSOM-based framework in such a challenging data regeneration task.[6]

The source data X is composed of seven basic log curves ($q = 7$). The target data Y is the image log. These two logs have to be adjusted due to their different depth acquisition rates[7], as illustrates Fig. 3.

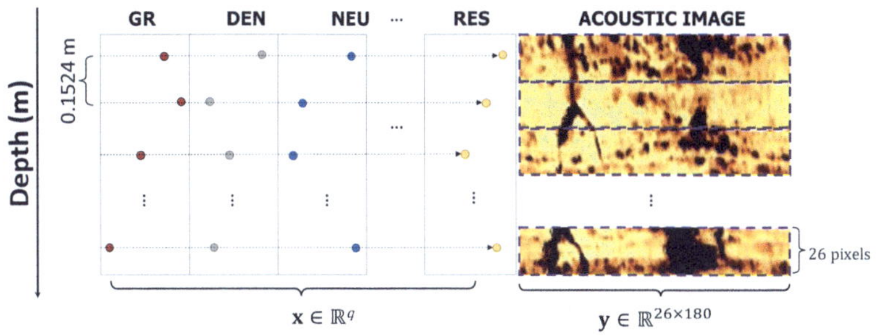

Fig. 3. Data depth adjustment to solve sampling rate differences. Each image slice ($\mathbf{y}$) is matched to a vector $\mathbf{x}$ of log values (coloured dots), centred at the base log depth.

3.1 Stages 1 and 3: Custom β-CVAE Model

Figure 4 (a) shows a β-CVAE model[8] used to reconstruct the target data and learn a reduced dimension space $\mathcal{Z}$ to represent the images (Stage 1). The decoder part, $d(.)$, is a generative model for acoustic images from $\mathbf{z}$ (Stage 3).

The quality of image generation from basic logs using the heteroassociative SOM mapping is limited by the quality of the image generated directly from the corresponding β-CVAE model fed with actual images. For this study, we focus on smaller dimensions for image representations. First, to lessen the possibility

---

[6] The reverse application can also be accomplished by our proposal (Stage 4). We could not explore it here due to page limits.

[7] An acoustic image has a new line every 0.0058 m, while basic logs have a new reading every 0.1524 m. Therefore, a 26×180 pixel image ($\mathbf{y}$) corresponds to a q-dimensional vector ($\mathbf{x}$) of log values.

[8] The β-CVAE model consists of three convolutional layers in the encoder and decoder, connected by two dense layers. The encoder has decreasing-size kernels (64, 32, 16 units) with 3×3 filters, while the decoder has increasing-size kernels (16, 32, 64 units) with 3×3 filters. The innermost layer (whose number of units is a hyperparameter) produces latent representations $\mathbf{z}$. Convolutional layers use SELU activation with He initialisation, without dropout or pooling layers. Fully connected layers are linear.

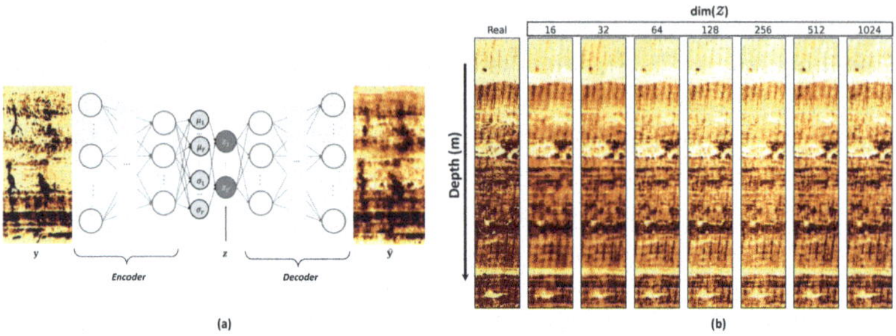

Fig. 4. (a) Illustration of a β-CVAE model for image representation learning and reconstruction. (b) Reconstructions by the different β-CVAE models, varying $dim(\mathcal{Z})$ (section of Well A).

that the technique will fail as a result of the dimensionality curse, which lowers the discriminability power of norms in higher dimensions. Second, to reduce the relative weight of the image component in $\mathbf{x}_{aug}$ when computing norms to determine the ERGSOM winners. As shown in Fig. 4 (b), even the low dimensional representations (e.g. $dim(\mathcal{Z}) = 16$) can regenerate reasonably acceptable images, particularly considering that such a regenerated image may come from a limited set of basic log curves.

3.2 Stage 2: ERGSOM-Based Methods

Based on the choice for incorporating the image information in $\mathbf{x}_{aug}$, we propose two methods to train ERGSOM for the heteroassociative mapping:

1. **ERGSOM** - using the β-CVAE's bottleneck layer to build the image representation:

$$\mathbf{x}_{aug} = [\mathbf{x}, \mathbf{z}] = [x_1, \ \ldots, \ x_q, \ z_1, \ \ldots, \ z_r] \tag{2}$$

2. p-**ERGSOM** - the representation is expressed in terms of the probabilistic information learnt by the β-CVAE model, as

$$\mathbf{x}_{aug} = [\mathbf{x}, \mu_z, \sigma_z] = [x_1, \ \ldots, \ x_q, \ \mu_1, \ \ldots, \ \mu_r, \sigma_1, \ \ldots, \ \sigma_r], \tag{3}$$

   in which the components of the vectors μ_z and σ_z are generated by the β-CVAE's encoder prior to the application of the reparametrization trick [5]. The weight vector for a unit i in the SOM becomes: $\mathbf{w}_i = [\mathbf{w}_i^x, \mathbf{w}_i^\mu, \mathbf{w}_i^\sigma]$.

Both variants of ERGSOM follow the same cross-modal inference procedure:

- Present as input to the trained ERGSOM only the basic logs vector $\mathbf{x}$, and find the winner i^* using just $\mathbf{w}_i^x$:

$$i^* = \arg\min_{i \in \mathcal{A}} \|\mathbf{x} - \mathbf{w}_i^x\| \, ; \tag{4}$$

– Retrieve the corresponding image representation $\mathbf{z}(\mathbf{x})$:
  • in the ERGSOM model, $\mathbf{z}(\mathbf{x}) = \mathbf{w}^z_{i*}$
  • in the p-ERGSOM model $\mathbf{z}(\mathbf{x}) = \mathbf{w}^{\mu}_{i*} + \mathbf{w}^{\sigma}_{i*} \odot \epsilon$, in which, $\epsilon \sim \mathcal{N}(\mathbf{0}, \mathbf{I}_m)$ and $\odot$ represents element-wise product.
– From $\mathbf{z}(\mathbf{x})$, generate the corresponding image using the decoder $d(.)$ of the β-CVAE model appropriate to the dimension of $\mathbf{z}$: $\hat{\mathbf{y}} = d(\mathbf{z})$.

Both approaches are compared for a section of Well A. For valid quantitative comparisons of the regenerated images, we rely on metrics used for image similarity assessment, namely MSE and SSIM. The MSE metric can sometimes diverge from human perceptual impressions, while the SSIM metric was designed to better fit human quality assessment [12].

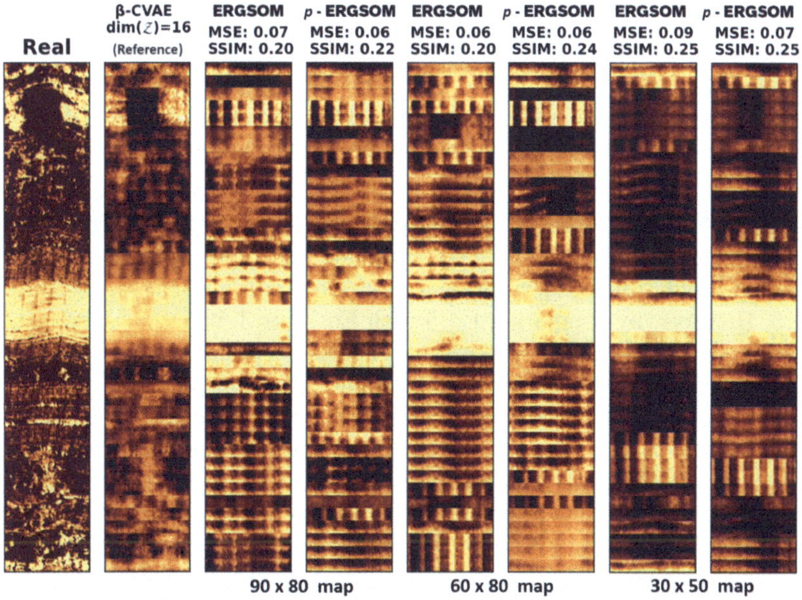

Fig. 5. Comparison of regenerated images: ERGSOM × p-ERGSOM (section of Well A). β-CVAE model with $\dim(\mathcal{Z}) = 16$ in the bottleneck layer.

Figure 5 shows the actual and the corresponding reference image generated by the β-CVAE model directly fed with it, followed by the regenerated images obtained by ERGSOM and p-ERGSOM[9]. We highlight two interesting results: 1) p-ERGSOM seems to increase images' similarity with the target image; 2) Smaller maps tend to deliver images closer to the target both, visually and quantitatively (e.g. SSIM = 0.25 for the p-ERGSOM image from the smaller

[9] Images are normalised to the [0,1] interval. The metrics use the β-CVAE image as reference.

30×50 map against a SSIM $= 0.20$ for the *p*-ERGSOM image from the 90×50 map). Larger maps seem to suffer more from the dissimilarity in the dimensions of the data components of $\mathbf{x}_{aug}$, specially for *p*-ERGSOM. Furthermore, we emphasise the smoother nature of the images regenerated by *p*-ERGSOM. It is noteworthy that even when generated by the same winning units, the adjacent images exhibit similarity but are never identical.[10].

Image Regeneration from Subsets of Basic Logs. ERGSOM-based models allow a type of inference not possible with other methods, especially those based on supervised learning: the regeneration of borehole image logs from a restricted subset of conventional petrophysical logs. This is accomplished as follows:

- Train ERGSOM with as many basic log curves as possible to compose $\mathbf{x}_{aug}$ (either using Eq. 2 or Eq. 3). During inference, the input $\mathbf{x}$ will be interpreted according to
$$\mathbf{x} = [\mathbf{x}_p, \mathbf{x}_m], \quad (5)$$
in which the index p means "present" and the index m means "missing";
- Once trained, present as input to ERGSOM only $\mathbf{x}_p$, and find ERGSOM's winner based on its units weight's component ($\mathbf{w}_{i*}^p$) corresponding to $\mathbf{x}_p$, and proceed as before.

Some authors (e.g. [11]) have applied the SOM in the limited scope of missing values imputation by implicitly assuming the input data $\mathbf{x}$ as a kind of augmented input composed by present and potentially missing values, just like in Eq. 5. However, by comparing Eq. 5 with Eqs. 2 or 3, it becomes evident that the use of the SOM for the imputation of missing values can be conceptually seen as a particular case of the proposed framework. Our methodology not only allows for the imputation of missing values, but also regenerates complex data from any subset of simpler inputs to the limit of a single input.

To illustrate, we apply the *p*-ERGSOM model in a section of Well A and show the regenerated images in Fig. 6[11]. The results of this experiment allow us to attest to the potential for recovering the most important information in the image from a reduced number of inputs. Visually, for subsets 6C and 5C, the larger structures, such as the dark structure at the top, and layers boundaries continue to be recovered. However, beyond this, the regenerated images start to drift and even show non-existent structures (subsets 3C and 2C). The use of a single input curve is merely illustrative of the potential of the framework, it is unlikely that such an extreme case would yield any valuable image[12].

[10] The data matching procedure depicted in Fig. 3 amplifies the "quantized" aspect of the reconstructed images, as each input produces a horizontal 26×180 image, which is vertically appended to the previous one to form the complete well section.

[11] We create seven curve subsets for this experiment: **1C** [DEN]; **2C** [DEN, NEU]; **3C** [GR, DEN, NEU]; **4C** [GR, DEN, NEU, log(RES)]; **5C** [DEN, NEU, DTC, nmrPhiT, PE]; **6C** [GR, DEN, NEU, DTC, PE, log(RES)] and **7C** with all curves.

[12] It is possible to achieve this by estimating the missing curves up to the total number of curves in the dataset, followed by using the full input (Eq. 2 or 3). Although promising, further investigation could not be conducted here due to page limits.

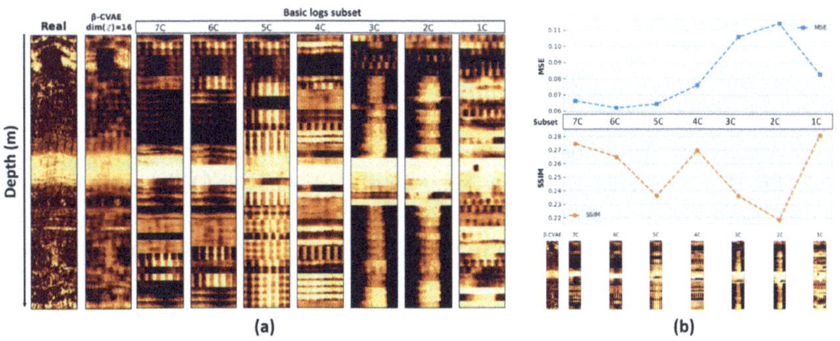

Fig. 6. (a) *p*-ERGSOM model images × Subset of input curves (section of Well A); (b) Image quality metrics for *p*-ERGSOM images, according to the subset of curves.

To better assess the quality of the regenerated images, we apply image quality metrics in Fig. 6 (b)[13]. While MSE appears to consistently worsen with the reduction in the number of input curves, SSIM reaches its maximum value for the subset with just one curve. However, despite this anomaly for Subset 1C, we can still match visual perception with the main trend in both metrics.

4 Discussion

The results corroborate the potential of the ERGSOM for image regeneration from basic logs, highlighting its power and flexibility. However, the application remains challenging due to the ill-posed nature of inferring complex images from limited raw measurements. Our method does not impose any constraints on the data relationship to guide the inference. Two factors may cause poor image regeneration:

1. **Data-related issues:** Retrieving 26×180 data points from a 7-dimensional input vector poses significant difficulties to any model, which should capture highly subtle data distribution and inter-relationship aspects. Although ERGSOM's results demonstrate promising prospects, the discrepancy in the dimensions of the components of $\mathbf{x}_{aug}$ negatively affects the SOM's training, especially in *p*-ERGSOM.
2. **Petrophysical issues:** Acoustic amplitude-based image logs are expected to be correlated only to basic logs related to the porosity of the rock. However, we input uncorrelated logs (such as GR and RES), expecting that this could help ERGSOM, which is a spatial mapping, by incorporating additional context information. Such an approach would be prohibitive in a supervised-based model as, in this case, only highly correlated inputs would be relevant to predict the output. Our method allows us to maximize the use of the data at hand.

[13] A quantitative analysis of the applied metrics in Fig. 6 can be misleading, which may suggest the need for more reliable metrics to deal with acoustic image logs.

5 Conclusions

This work introduces a fully unsupervised framework using the SOM to learn cross-modal representations. We extend the classic SOM-based associative memory technique to a heteroassociative source-to-target mapping, aligning it with the Regenerative Learning paradigm. The ERGSOM-based framework generates complex data from simpler inputs using cross-domain mapping. The framework was successfully tested in a challenging petrophysics application, generating coherent image logs from basic log curves. It proved capable of generating such images from any number of input curves, even with a single dimensional input component. The ERGSOM is a powerful and flexible framework adaptable for various inter-domain prediction or data generation applications.

Acknowledgements. The authors acknowledge PETROBRAS for permission to publish this paper. This study was financed in part by the Brazilian Agencies CAPES (Finance Code 001), CNPq (grants 304497/2017-7 and 313251/2023-1) and FAPERJ (E-26/201.110/2022).

References

1. Barreto, G., Araujo, A.: Identification and control of dynamical systems using the self-organizing map. IEEE TNN **15**(5), 1244–1259 (2004)
2. Barreto, G., Araujo, A., Ritter, H.: Self-organizing feature maps for modeling and control of robotic manipulators. JIRS **36**(4), 407–450 (2003)
3. de Gouvêa, R.C.T., Gioria, R., Marques, G.R., Carneiro, C.C.: IntraSOM: a comprehensive python library for self-organizing maps with hexagonal toroidal maps training and missing data handling. Softw. Impacts **17**, 100570 (2023)
4. Higgins, I., Matthey, L., et al.: β-VAE: learning basic visual concepts with a constrained variational framework. In: ICLR (2017)
5. Kingma, D.P., Welling, M.: Auto-encoding variational bayes. arXiv preprint arXiv:1312.6114 (2013)
6. Kohonen, T.: Self-organizing Maps, 3rd edn. Springer, Heidelberg (2001). https://doi.org/10.1007/978-3-642-56927-2
7. Niitsuma, J., Osana, Y.: Self-organizing map-based probabilistic associative memory for sequential patterns. In: IJCNN 2015, pp. 1–7 (2015)
8. Ritter, H., Schulten, K.: Topology conserving mappings for learning motor tasks. In: AIP Conference Proceedings, vol. 151, pp. 376–380. AIP (1986)
9. Tan, X., Qin, T., Bian, J., Liu, T.Y., Bengio, Y.: Regeneration learning: a learning paradigm for data generation. arXiv preprint arXiv:2301.08846 (2023)
10. Tiab, D., Donaldson, E.C.: Petrophysics. Elsevier, Amsterdam (2016)
11. Wang, S.: Application of self-organising maps for data mining with incomplete data sets. Neural Comput. Appl. **12**, 42–48 (2003)
12. Wang, Z., Bovik, A., Sheikh, H., Simoncelli, E.: Image quality assessment: from error visibility to structural similarity. IEEE TIP **13**(4), 600–612 (2004)
13. Yao, Y., Liu, Z., Lin, Y., Sun, M.: Cross-modal representation learning. In: Liu, Z., Lin, Y., Sun, M. (eds.) Representation Learning for Natural Language Processing, pp. 211–240. Springer, Singapore (2023). https://doi.org/10.1007/978-981-99-1600-9_7

Unsupervised Learning-Based Data Collection Planning with Dubins Vehicle and Constrained Data Retrieving Time

Jindřiška Deckerová(✉) and Jan Faigl

Faculty of Electrical Engineering, Czech Technical University in Prague, Technická 2, 166 27 Prague, Czechia
{deckejin,faiglj}@fel.cvut.cz

Abstract. In remote data collection from sampling stations, a vehicle must be within sufficient distance from a particular station for a predefined minimal time to retrieve required data from the site. The planning task is to find a cost-efficient data collection plan to retrieve data from all the stations. For a fixed-wing aerial vehicle flying with a constant forward velocity, the problem is to determine the shortest feasible path that visits every sensing site and ensure the vehicle is within a reliable communication distance from the station for a sufficient period. We propose to formulate the planning problem as a variant of the Close Enough Dubins Traveling Salesman Problem with Time Constraints (CEDTSP-TC) that is heuristically solved by unsupervised learning of the Growing Self-Organizing Array (GSOA) modified to address the constrained minimal data retrieving time. The proposed method is compared with a baseline based on a sampling-based decoupled approach, and the results support the feasibility of both proposed solvers in random instances.

Keyword: Unsupervised Learning, Growing Self-Organizing Array, Data Collection, Traveling Salesman Problem, Dubins Vehicle

1 Introduction

The studied data collection planning is motivated by the tasks where fixed-wing aerial vehicles retrieve data from sampling stations using wireless communication. The data retrieval is possible only when the vehicle is within a sufficient distance from a sampling station. Besides, the vehicle must be within the distance for a predefined minimal time to retrieve data with a limited data transfer rate. The data collection planning problem with the fixed-wing's motion constraints on the minimal flying velocity and limited vehicle turning radius can be formulated as the *Dubins Traveling Salesman Problem* (DTSP) [22], where the travel cost from one location to another location corresponds to the length of the shortest curvature-constrained (Dubins) path.

Dubins path makes the DTSP a combination of combinatorial and continuous optimization problems to determine the optimal sequence of visits to the

T. Villmann et al. (Eds.): WSOM+2024, LNNS 1087, pp. 11–21, 2024.
https://doi.org/10.1007/978-3-031-67159-3_2

target locations and determine the vehicle's heading angles at the locations since the length of the shortest path depends on the angles. Because of the underlying combinatorial TSP, the DTSP is also NP-hard [19]. The DTSP has been addressed by several approaches, such as [1,4,13,17,21,22,26]. Challenging sequence-dependent routing can be addressed by a decoupled approach, where the sequence of visits is determined for the relaxed heading constraints. Then, optimal headings are determined for the sequence using the Dubins path connecting the locations. Besides, the problem can be discretized into purely combinatorial problems by sampling possible heading angles into some finite locations.

In the motivational data collection, the communication range can be exploited in finding the optimal solution of visiting disk-shaped regions around the sampling stations formulated as the *DTSP with Neighborhoods* [20]. Here, in addition to the optimal vehicle heading angle at the data retrieval location, we search for the optimal locations of visits to the regions (neighborhoods). Similar to DTSP, decoupled [15,17,22,25], sampling-based [2], and direct approaches [11,12,16,18,26] have been proposed.

Although the DTSPN might yield shorter tours than using the DTSP with the centers of the disk-shaped neighborhoods, the minimal required time the vehicle spent within the station's communication range is not guaranteed. Therefore, a novel problem formulation is proposed to enable data retrieval by an uninterrupted vehicle presence within the sampling station's neighborhood for the defined minimal time. The introduced problem is studied as the *Close Enough DTSP with Time Constraints* (CEDTSP-TC) to highlight the disk-shaped neighborhoods. The problem turns into determining two waypoint locations (entering and leaving) for each region (disk), such that the waypoints are connected by a minimal length path entirely within the region to retrieve all data.

We propose to address the introduced CEDTSP-TC by unsupervised learning of the Growing Self-Organizing Array (GSOA) [6] already successfully deployed in various data collection planning problems as variants of the TSP [3,8–10] providing competitive solutions to existing heuristic methods in short computational times. The GSOA simultaneously determines the sequence of visits with the waypoints and headings [4] and can be relatively easy to modify to other problems, such as addressing spatially correlated measurements [7]. For the CEDTSP-TC, we need to address the minimal length path within the region.

The performance of the developed GSOA-based solution is compared with a baseline decoupled sampling-based approach Centroid-TSP, where the sequence is determined as a solution of the Euclidean TSP (ETSP) using the centers of the disk-shaped regions; then, waypoints are determined from a set of sampled locations on the border of the regions, each with a set of sampled possible heading angles. Both approaches provide competitive results, but GSOA outperforms the baseline method in computational time and solving instances with a low number of samples.

The rest of the paper is organized as follows. The problem is formally defined in Sect. 2. The proposed Centroid-TSP- and GSOA-based approaches are described in Sect. 3 and Sect. 4, respectively. The empirical evaluation is presented in Sect. 5, and concluding remarks can be found in Sect. 6.

2 Problem Statement

The studied CEDTSP-TC is to find the cost-efficient curvature-constrained path to visit a given set of disk-shaped regions such that the path passes the regions with at least $\mathcal{L}_{\min}$ length to remotely collect data from the sampling station located at the center of the disk. The curvature-constrained path consists of a sequence of Dubins paths feasible for Dubins vehicle that moves forward with a constant velocity v and has limited minimal turning radius ρ [5]. The vehicle motion can be described by a state equation for a control input u as

$$\begin{bmatrix} \dot{x} \\ \dot{y} \\ \dot{\theta} \end{bmatrix} = v \begin{bmatrix} \cos\theta \\ \sin\theta \\ \frac{u}{\rho} \end{bmatrix}, \quad |u| \leq 1, \tag{1}$$

where $(x, y) \in \mathbb{R}^2$ is the vehicle's position in the plane and $\theta \in (0, 2\pi]$ denotes the vehicle's heading angle at (x, y). The state of the vehicle is thus $\boldsymbol{q} = (x, y; \theta)$ and $\boldsymbol{q}$ is from the Special Euclidean group $\boldsymbol{q} \in SE(2)$.

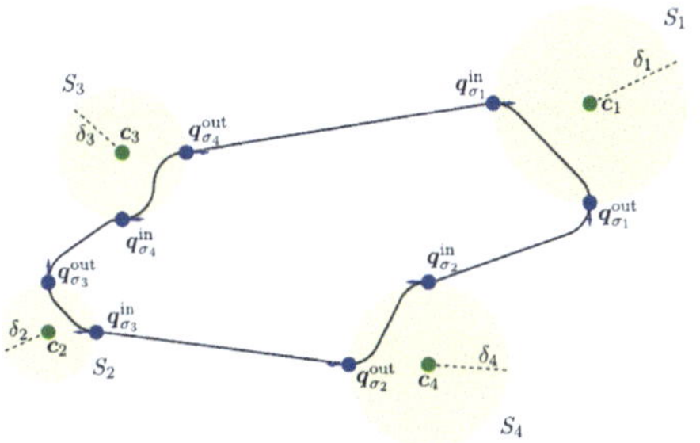

(a) Example of the CEDTSP-TC instance with the used notation.

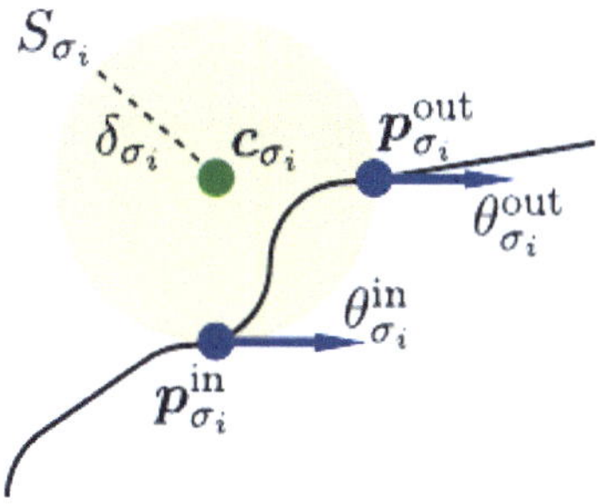

(b) Dubins data collection path path in the disk S_{σ_i}.

Fig. 1. An instance of the CEDTSP-TC with $n = 4$ target regions $\mathcal{S}$ depicted in yellow (a) together with its solution. The solution is the curvature-constrained path (black) visiting $\mathcal{S}$ in the order $\Sigma = (1, 4, 2, 3)$ at the configurations $Q = \{\boldsymbol{q}^{\text{in}}_{\sigma_1}, \boldsymbol{q}^{\text{out}}_{\sigma_1}, \ldots, \boldsymbol{q}^{\text{in}}_{\sigma_n}, \boldsymbol{q}^{\text{out}}_{\sigma_n}\}$ (blue). The right subfigure (b) shows a detail of Dubins path within a region with the input configuration $\boldsymbol{q}^{\text{in}}_{\sigma_i} = (\boldsymbol{q}^{\text{in}}_{\sigma_i}; \theta^{\text{in}}_{\sigma_i})$ and output configuration $\boldsymbol{q}^{\text{out}}_{\sigma_i} = (\boldsymbol{q}^{\text{out}}_{\sigma_i}; \theta^{\text{out}}_{\sigma_i})$ that consists of points of visits $\boldsymbol{p}^{\text{in}}_{\sigma_i}$ and $\boldsymbol{p}^{\text{out}}_{\sigma_i}$ and corresponding vehicle heading angles $\theta^{\text{in}}_{\sigma_i}$ and $\theta^{\text{out}}_{\sigma_i}$, respectively.

Dubins path between two configurations $\boldsymbol{q}_i$ and $\boldsymbol{q}_j$ is the shortest path satisfying constraints of Dubins vehicle (1). The path can be found as a closed-form solution [5], and it consists of up to three segments of a straight line (S) and circular arc (C) with the radius ρ. The arc can be a left (L) and right (R) turn that gives up six possible combinations: LSL, LSR, RSL, RSR, LRL, and RLR. The length of Dubins path is denoted $\mathcal{L}(\boldsymbol{q}_i, \boldsymbol{q}_j)$.

The CEDTSP-TC formulation follows the previous work [8] and is to visit a given set of n disk-shaped regions $\mathcal{S} = \{S_1, \ldots, S_n\}$, each $S_i \in \mathbb{R}^2$ defined by its center $c_i \in \mathbb{R}^2$ and radius $\delta_i > 0$. However, the data collection path is requested to pass each region's S_i with the minimal length $\mathcal{L}_{\min}$, and such a part of the path is entirely with the region. Then, the solution of the CEDTSP-TC is a sequence $\Sigma = (\sigma_1, \ldots, \sigma_n)$, $\sigma_i \neq \sigma_j$, for $i \neq j$, of visits to $\mathcal{S}$ and a set of configurations Q denoting entering and leaving configurations of the path to visit each region. For a region S_i, the entering configuration is denoted $\boldsymbol{q}_i^{\text{in}}$ and leaving configuration $\boldsymbol{q}_i^{\text{out}}$. The path between the entering and leaving configurations is at least $\mathcal{L}_{\min}$ long, $\mathcal{L}(\boldsymbol{q}_i^{\text{in}}, \boldsymbol{q}_i^{\text{out}}) \geq \mathcal{L}_{\min}$. The set Q can be defined as $Q = \{\boldsymbol{q}_1^{\text{in}}, \boldsymbol{q}_1^{\text{out}}, \ldots, \boldsymbol{q}_n^{\text{in}}, \boldsymbol{q}_n^{\text{out}}\}$. CEDTSP-TC is formally defined as Problem 1, and a problem instance with the used notation is depicted in Fig. 1.

Problem 1. (Close Enough DTSP with Time Constraints (CEDTSP-TC)).

$$\underset{\Sigma, Q}{\text{minimize}} \quad \mathcal{L}(\boldsymbol{q}_{\sigma_n}^{\text{in}}, \boldsymbol{q}_{\sigma_n}^{\text{out}}) + \mathcal{L}(\boldsymbol{q}_{\sigma_n}^{\text{out}}, \boldsymbol{q}_{\sigma_1}^{\text{in}}) + \sum_{i=1}^{n-1} \Big(\mathcal{L}(\boldsymbol{q}_{\sigma_i}^{\text{in}}, \boldsymbol{q}_{\sigma_i}^{\text{out}}) + \mathcal{L}(\boldsymbol{q}_{\sigma_i}^{\text{out}}, \boldsymbol{q}_{\sigma_{i+1}}^{\text{in}}) \Big) \tag{2}$$

$$\text{s.t.} \quad \Sigma = (\sigma_1, \ldots, \sigma_n),\ 1 \leq \sigma_i \leq n,\ \sigma_i \neq \sigma_j \text{ for } i \neq j \tag{3}$$

$$Q = \{\boldsymbol{q}_1^{\text{in}}, \boldsymbol{q}_1^{\text{out}}, \ldots, \boldsymbol{q}_n^{\text{in}}, \boldsymbol{q}_n^{\text{out}}\} \tag{4}$$

$$\mathcal{L}(\boldsymbol{q}_i^{\text{in}}, \boldsymbol{q}_i^{\text{out}}) \geq \mathcal{L}_{\min},\ 1 \leq i \leq n \tag{5}$$

$$\boldsymbol{q}_i^{\text{in}}, \boldsymbol{q}_i^{\text{out}} \in SE(2),\ 1 \leq i \leq n \tag{6}$$

$$\boldsymbol{q}_i^{\text{in}} = (x_i^{\text{in}}, y_i^{\text{in}}, \theta_i^{\text{in}}),\ \boldsymbol{q}_i^{\text{out}} = (x_i^{\text{out}}, y_i^{\text{out}}, \theta_i^{\text{out}}), \tag{7}$$

$$\left\|(x_i^{\text{in}}, y_i^{\text{in}}) - \boldsymbol{c}_i\right\| \leq \delta_i,\ \left\|(x_i^{\text{out}}, y_i^{\text{out}}) - \boldsymbol{c}_i\right\| \leq \delta_i,\ 1 \leq i \leq n \tag{8}$$

3 Decoupled Sampling-Based Solution

The decoupled sampling-based approach Centroid-TSP decomposes the problem into the (i) determination of sequence Σ and (ii) configurations Q. The sequence of visits Σ to $\mathcal{S}$ is determined from a solution of the ETSP instance with centers of the regions and Euclidean distance connecting them, for example, using the available GLKH solver [14]. The configurations Q are determined by sampling each S_i at the region border uniformly to ω locations, and for each location, ω heading angles are uniformly sampled. Thus, each S_i is sampled into $\omega \times \omega$ configurations $\mathbf{Q}_i = \{\boldsymbol{q}_1^i, \ldots, \boldsymbol{q}_{\omega \times \omega}^i\}$. The number of samples ω denotes the sampling resolution, and increasing ω might lead to an improved solution.

For a sequence Σ and a set of sampled configurations Q, an oriented search graph with $2n$ layers can be constructed as depicted in Fig. 2. The edges between the nodes of consecutive layers denote Dubins path connecting the associated configurations with the cost corresponding to the length of the path. If the path connecting the entering and leaving configuration of the region is shorter than $\mathcal{L}_{\min}$, the edge does not connect the corresponding nodes, and its cost is set to ∞. The particular solution Q is found by finding the shortest path in the graph using the dynamic program, whose complexity can be bounded by $O(2n\omega^6)$ [24]. Hence, the solution of the CEDTSP-TC is determined as a solution (Σ, Q) of its discretized variant using ω^2 samples.

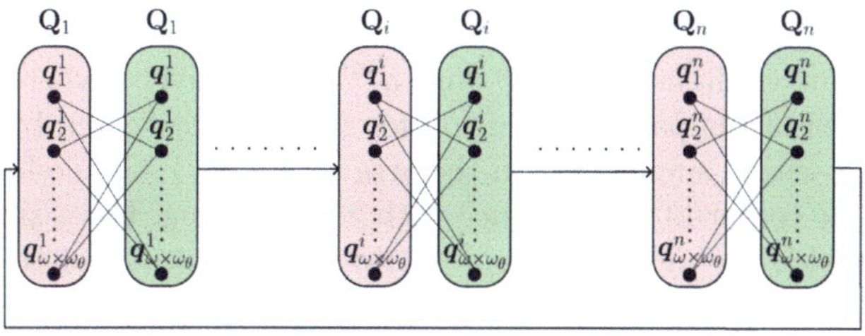

Fig. 2. An oriented search graph with $2n$ layers for n target regions of the CEDTSP-TC. The red layers correspond to the regions' input configurations and the green layers to the output configurations.

4 Growing Self-organizing Array for the CEDTSP-TC

The proposed method to the CEDTSP-TC is based on adopting the unsupervised learning-based heuristic for routing problems GSOA [6], already applied to Dubins routing problems in [8]. Therefore, only the GSOA summary is presented with a focus on the modifications needed to address the CEDTSP-TC. Two variants of the GSOA applications to address the non-Euclidean problem are proposed to demonstrate further prospective modifications. The learning procedure is summarized in Algorithm 1 and works as follows.

Algorithm 1: GSOA for the CEDTSP-TC

Input: $S = \{S_1, \ldots, S_n\}$ – a set of regions, $\mathcal{L}_{\min}$ – minimal path length.
Params. : $G = 10$ – the learning gain; $\alpha = 0.0005$ – the gain decreasing rate; $\mu = 0.6$ – the learning rate; and $i_{\max} = \min(120, 1/\alpha)$ – the no. of learning epochs; that are set as suggested in [6].
Output: (Σ, Q) – found a solution.

```
N ← {ν_1}, i ← 0                    // Initialize with geometric center
while i ≤ i_max do
    foreach S_i in random permutation of S do
        Select winner node ν* of N for S_i
        Insert ν* into N
        foreach ν ∈ N with d distance from ν*, where 0 ≤ d ≤ 0.2M do
            Adapt ν toward ν*.s_p according to (10) with (11)
    Remove nodes from N from the previous epoch
    i ← i + 1, G ← (1 − iα)G
    Construct (Σ', Q') by traversing N from the associated q_i^in and q_i^out
    (Σ, Q) ← (Σ', Q') if (Σ', Q') is shorter than (Σ, Q)
return (Σ, Q)
```

The GSOA is an iterative procedure where the solution is encoded as an array of nodes $\mathcal{N} = (\nu_1, \ldots, \nu_M)$. Each node ν_j is associated with its location $\boldsymbol{\nu}_j \in \mathbb{R}^2$ and heading $\theta_s \in [0, 2\pi)$, the target region $S_{\sigma_i} \in \mathcal{S}$, the waypoint location

$\boldsymbol{s}_p \in \mathbb{R}^2$, $\|\boldsymbol{s}_p - \boldsymbol{c}_{\sigma_i}\| \leq \delta_{\sigma_i}$, at which the target is visited, and further with the input and output configurations $\boldsymbol{q}^{\text{in}}_{\sigma_i}$ and $\boldsymbol{q}^{\text{out}}_{\sigma_i}$, respectively, where $1 \leq \sigma_i \leq n$ denotes the label of the particular region for the sequence given by the array of nodes. Starting with a single node in the array $\mathcal{N} = \{\nu_1\}$, the node is initialized to the geometric center of the instance with zero heading. For a single iteration (a learning epoch), new nodes are added into $\mathcal{N}$ for each target that is selected randomly from $\mathcal{S}$ to avoid location minima. The node is added as the most suitable for the particular target (the closest point of the path represented by $\mathcal{N}$) and adapted toward the region. The added nodes have associated regions and locations of visits. Therefore, all the previous nodes can be removed from $\mathcal{N}$, and a solution can be retrieved from the array after every epoch. The adaptation can be repeated until the maximum number of iterations $i_{\max}$ is reached.

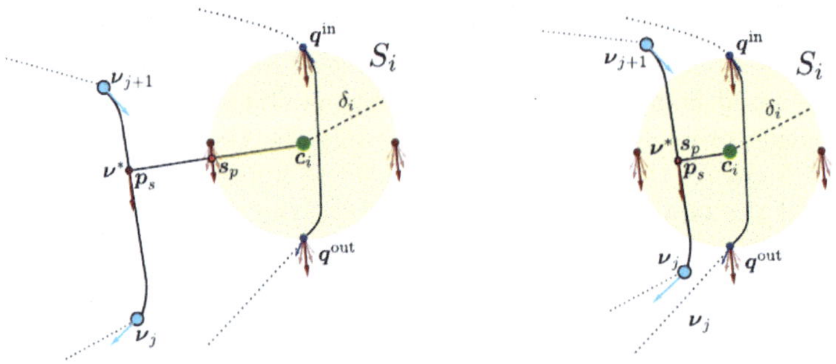

(a) Winner node $\boldsymbol{\nu}^*$ outside of S_i. (b) Winner node $\boldsymbol{\nu}^*$ inside of S_i.

Fig. 3. Determination of the winner node ν^* for the target region S_i at $\boldsymbol{\nu}^*$ with the points $\boldsymbol{p}_s$ and $\boldsymbol{s}_p$, and configurations $\boldsymbol{q}^{\text{in}}$ and $\boldsymbol{q}^{\text{out}}$. Dubins path between ν_j and ν_{j+1} can be used to sample the heading angles during the adaptation.

A new node added to $\mathcal{N}$ is called the *winner node* ν^* as it is determined for the particular target S_i as the closest point of the path represented by $\mathcal{N}$ to S_i as visualized in Fig. 3. Approximation of Dubins tour can be constructed from the array $\mathcal{N}$ with M nodes using two consecutive nodes of $\mathcal{N}$ to determine Dubins path (ν_j, ν_{j+1}), $\nu_{M+1} = \nu_1$. By iterating over all such Dubins paths, we can determine $\boldsymbol{p}_s$ together with the corresponding heading θ_s of the particular Dubins path as the closest point to S_i centered at $\boldsymbol{c}_i$. The location of the new node ν^* is set to $\boldsymbol{p}_s$, and its heading can be θ_s. The node ν^* is added to $\mathcal{N}$ between the corresponding nodes ν_j and ν_{j+1}, and declared to be the current winner node. Besides, if $\boldsymbol{p}_s$ is inside S_i, we can set the associated waypoint $\boldsymbol{s}_p$ to $\boldsymbol{p}_s$. Otherwise, $\boldsymbol{s}_p$ of the data collection path visits to S_i is determined as the intersection of S_i border and segment $(\boldsymbol{p}_s, \boldsymbol{c}_i)$ as depicted in Fig. 3a.

Since each region needs to be visited with at least $\mathcal{L}_{\min}$ length path entirely in the region, the input and output configurations $\boldsymbol{q}^{\text{in}}_i$ and $\boldsymbol{q}^{\text{out}}_i$, respectively, need to be determined. For the initial employment of GSOA, denoted GSOA-naive, we can utilize a similar approach to the CENTROID-TSP and examine sampled configurations $Q = \{\boldsymbol{q}_1, \ldots, \boldsymbol{q}_{\omega\times\omega}\}$, where ω denotes the number of sampled headings at each possible input and output locations. For simplicity, we can utilize the uniform samples as in Sect. 3. Then, q_i^{in} and q_i^{out} are determined as the pair that minimizes the path length from the node ν_j to ν_{j+1} as

$$q^{\text{in}}, q^{\text{out}} = \underset{\substack{q_1, q_2 \in Q \\ \mathcal{L}(q_1, q_2) \geq \mathcal{L}_{\min}}}{\arg\min} \quad \mathcal{L}(\nu_j.q^{\text{out}}, q_1) + \mathcal{L}(q_1, q_2) + \mathcal{L}(q_2, \nu_{j+1}.q^{\text{in}}), \qquad (9)$$

where $\mathcal{L}(\cdot, \cdot)$ denotes the length of Dubins path, and the path from q_1 to q_2 is entirely inside the region, and its length is at least $\mathcal{L}_{\min}$.

In the naive approach of GSOA-naive, Q is uniformly sampled. However, we can exploit the exploration capability of GSOA by using the determined heading angle θ_s of the winner node and sample k heading angles locally in the sampling interval $(\theta_s - \varepsilon, \theta_s + \varepsilon)$, where ε denotes the sampling range, and $k \leq \omega$, see Fig. 3. Besides, θ_s is determined as the vehicle heading angle at the midpoint of Dubins path from ν_j to ν_{j+1}. Then, q_i^{in} and q_i^{out} are determined according to (9).

Once the winner node ν^* is added into $\mathcal{N}$, it is adapted with its neighboring nodes that are no more than $0.2M$ nodes apart toward S_i, but only if p_s is not already inside of S_i, i.e., $p_s \neq s_p$, see Fig. 3b. Note that the size of the neighboring nodes $0.2M$ denotes the activation bubble employed in one of the earliest unsupervised learning approaches to the TSP [23]. The adaptation adjusts the node's location ν toward the waypoint s_p of ν^* as of

$$\nu \leftarrow \nu + \mu f(G, d)(p_s - \nu), \qquad (10)$$

where μ is the learning rate, G is the learning gain, d is the number of nodes between the adapted node ν and winner node ν^*. The neighboring function $f(G, d)$ [12] influences the power of adaptation as of

$$f(G, d) = \begin{cases} e^{-\frac{d^2}{G^2}} & \text{if } d < 0.2M \\ 0 & \text{otherwise} \end{cases}. \qquad (11)$$

Since n new nodes are added to $\mathcal{N}$ in each learning epoch, the nodes from the previous epoch are removed. Besides, the learning rate is updated to $\mu \leftarrow (1 - i\alpha)\mu$, where α is the gain decreasing rate, and i is the current number of the performed learning epoch. A solution of the CEDTSP-TC can be obtained after each learning epoch by traversing the array $\mathcal{N}$ and using the associated q^{in} and q^{out}. The best solution found so far can be maintained during the learning.

We follow the parameterization of GSOA as reported in [6] that is selected based on an empirical evaluation. The learning gain G is set to 10, the gain decreasing rate α is set to 0.0005, and the learning rate μ is set to 0.6. The maximal number of learning epochs is set as the smaller value of $i_{\max}$ ($i_{\max} = 120$) or $1/\alpha$ to ensure G is always above zero.

5 Empirical Evaluation

The proposed solvers (Centroid-TSP and GSOA) to the introduced CEDTSP-TC have been empirically studied in solving 30 randomly generated instances with $\delta \geq \mathcal{L}_{\min}$. Each instance is denoted **random_**n**_**$[a-e]$, where n corresponds to the number of target regions $n \in \{5, 6, 7, 8, 9, 10\}$, and $[a-e]$ denotes particular instances with the same n. The turning radius is set to $\rho = 1.0$, the resolution ω to $\{2, 4, 8, 16, 32\}$, the minimal required length to $\mathcal{L}_{\min} = 0.5$, such as there exists at least one configuration that follows (5). Besides, an ablation study is performed by comparing the performance of the GSOA with that of the GSOA-naive to show the benefits of online sampling.

The solution quality is measured using *relative length* $\mathcal{L}_{\text{rel}}$ of the solution cost found by a particular solver for a particular instance, $\mathcal{L}_{\text{rel}} = (\mathcal{L}/\mathcal{L}^*)$, where $\mathcal{L}^*$ is the best solution cost found among all examined solvers for the particular instance, and $\mathcal{L}$ is the best solution length found for the particular instance among all performed trials of the particular solver. The Centroid-TSP solver employs the GLKH solver [14]. Since the GSOA is a stochastic algorithm, the number of performed trials per instance is set to 10. The number of heading angle samples in the GSOA-naive is ω as in the Centroid-TSP. However, the GSOA uses $k = 4$ samples if $\omega < 8$, otherwise $k = 16$, for the sampling range $\varepsilon = \pi/8$ if $\omega < 8$, otherwise $\varepsilon = \pi/2$.

The solvers are executed using the `Intel® Core™ i9-13900K` and the computational requirements are reported as the average required computational time $\mathrm{T}_{\mathrm{CPU}}$ (in seconds) to obtain a single solution of a particular instance by the solver. A summary of the results is depicted in Fig. 4.

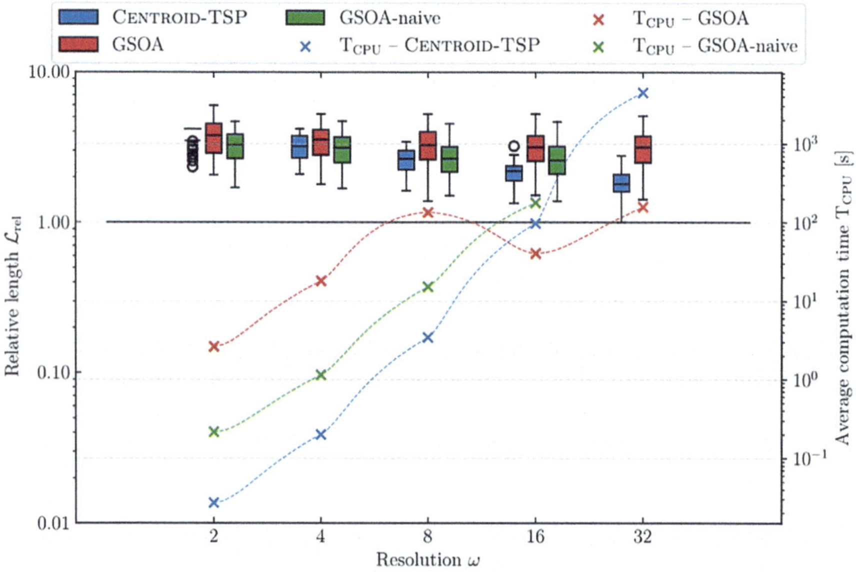

Fig. 4. Effect of resolution ω on the solution length $\mathcal{L}_{\text{rel}}$ and computation time $\mathrm{T}_{\mathrm{CPU}}$ [s] depicted as the five-point summary.

The results indicate that the performance of the GSOA is competitive with the Centroid-TSP. For solving instances with low resolutions ω, the GSOA is more demanding than Centroid-TSP; however, GSOA benefits from online sampling in comparison to GSOA-naive and Centroid-TSP for a dense resolution above 16. Possible solution improvements of the GSOA compared to the GSOA-naive are not significant, but GSOA-naive fails to find solutions for $\omega > 16$, while GSOA is less demanding, which might be further improved.

6 Conclusion

A novel data collection planning problem with Dubins vehicle has been introduced as the Close Enough DTSP with Time Constraints (CEDTSP-TC). It is motivated by the practical need to retrieve data with limited data transfer speed, which requires the data collection vehicle to be within the communication range for a minimum data collection time. Two solution approaches are proposed: the decoupled sampling-based method and unsupervised learning of the GSOA. The reported results support the feasibility of both approaches. Although the developed GSOA heuristic provides competitive results with less computational requirements than the decoupled method, the ablation study indicates further improvements might be possible, which is a subject of our future work.

Acknowledgments. The work was supported by the Czech Science Foundation (GAČR) under research project No. 22-05762S.

References

1. Babel, L.: New heuristic algorithms for the Dubins traveling salesman problem. J. Heurist. **26**(4), 503–530 (2020). https://doi.org/10.1007/s10732-020-09440-2
2. Cohen, I., Epstein, C., Isaiah, P., Kuzi, S., Shima, T.: Discretization-based and look-ahead algorithms for the Dubins traveling salesperson problem. IEEE Trans. Autom. Sci. Eng. **14**(1), 383–390 (2017). https://doi.org/10.1109/TASE.2016.2602385
3. Deckerová, J., Faigl, J., Krátký, V.: Traveling salesman problem with neighborhoods on a sphere in reflectance transformation imaging scenarios. Expert Syst. Appl. **198**, 116814 (2022). https://doi.org/10.1016/j.eswa.2022.116814
4. Drchal, J., Faigl, J., Váňa, P.: WiSM: windowing surrogate model for evaluation of curvature-constrained tours with Dubins vehicle. IEEE Trans. Cybern. **52**(2), 1302–1311 (2022). https://doi.org/10.1109/TCYB.2020.3000465
5. Dubins, L.E.: On curves of minimal length with a constraint on average curvature, and with prescribed initial and terminal positions and tangents. Am. J. Math. **79**, 497 (1957). https://doi.org/10.2307/2372560
6. Faigl, J.: GSOA: growing self-organizing array - unsupervised learning for the close-enough traveling salesman problem and other routing problems. Neurocomputing **312**, 120–134 (2018). https://doi.org/10.1016/j.neucom.2018.05.079
7. Faigl, J.: Data collection path planning with spatially correlated measurements using growing self-organizing array. Appl. Soft Comput. **75**, 130–147 (2019). https://doi.org/10.1016/j.asoc.2018.11.005
8. Faigl, J.: Unsupervised learning-based solution of the close enough Dubins orienteering problem. Neural Comput. Appl. **24**(32), 18193–18211 (2020). https://doi.org/10.1007/s00521-019-04222-9

9. Faigl, J., Deckerová, J.: On unsupervised learning based multi-goal path planning for visiting 3D regions. In: International Conference on Robotics and Artificial Intelligence (ICRAI), pp. 45–50 (2018). https://doi.org/10.1145/3297097.3297099
10. Faigl, J., Deckerová, J., Váňa, P.: Fast heuristics for the 3D multi-goal path planning based on the generalized traveling salesman problem with neighborhoods. IEEE Robot. Autom. Lett. **4**, 2439–2446 (2019). https://doi.org/10.1109/LRA.2019.2900507
11. Faigl, J., Váňa, P.: Self-organizing map for the curvature-constrained traveling salesman problem. In: Villa, A.E.P., Masulli, P., Pons Rivero, A.J. (eds.) ICANN 2016. LNCS, vol. 9887, pp. 497–505. Springer, Cham (2016). https://doi.org/10.1007/978-3-319-44781-0_59
12. Faigl, J., Váňa, P.: Unsupervised learning for surveillance planning with team of aerial vehicles. In: International Joint Conference on Neural Networks (IJCNN), pp. 4340–4347 (2017). https://doi.org/10.1109/IJCNN.2017.7966405
13. Faigl, J., Váňa, P., Drchal, J.: Fast sequence rejection for multi-goal planning with dubins vehicle. In: IEEE/RSJ International Conference on Intelligent Robots and Systems (IROS), pp. 6773–6780 (2020). https://doi.org/10.1109/IROS45743.2020.9340644
14. Helsgaun, K.: GLKH (2013). http://akira.ruc.dk/~keld/research/GLKH/. Accessed 8 Jan 2024
15. Isaacs, J.T., Klein, D.J., Hespanha, J.P.: Algorithms for the traveling salesman problem with neighborhoods involving a Dubins vehicle. In: American Control Conference, pp. 1704–1709 (2011). https://doi.org/10.1109/ACC.2011.5991501
16. Macharet, D.G., Campos, M.F.M.: A survey on routing problems and robotic systems. Robotica **36**(12), 1781–1803 (2018). https://doi.org/10.1017/S0263574718000735
17. Macharet, D.G., Neto, A.A., da Camara Neto, V.F., Campos, M.F.: Nonholonomic path planning optimization for Dubins' vehicles. In: International Conference on Robotics and Automation (ICRA), pp. 4208–4213 (2011). https://doi.org/10.1109/ICRA.2011.5980239
18. Macharet, D.G., Neto, A.A., da Camara Neto, V.F., Campos, M.F.: Data gathering tour optimization for Dubins' vehicles. In: Congress on Evolutionary Computation (CEC), pp. 1–8. IEEE (2012). https://doi.org/10.1109/CEC.2012.6256477
19. Ny, J., Feron, E., Frazzoli, E.: On the Dubins traveling salesman problem. IEEE Trans. Autom. Control **57**(1), 265–270 (2012). https://doi.org/10.1109/TAC.2011.2166311
20. Obermeyer, K.: Path planning for a UAV performing reconnaissance of static ground targets in terrain. In: 2009 AIAA Guidance, Navigation, and Control Conference, pp. 1–11 (2012). https://doi.org/10.2514/6.2009-5888
21. Obermeyer, K., Oberlin, P., Darbha, S.: Sampling-based roadmap methods for a visual reconnaissance UAV. In: AIAA Guidance, Navigation, and Control Conference, p. 7568 (2010). https://doi.org/10.2514/6.2010-7568
22. Savla, K., Frazzoli, E., Bullo, F.: On the point-to-point and traveling salesperson problems for Dubins' vehicle. In: American Control Conference, vol. 2, pp. 786–791 (2005). https://doi.org/10.1109/ACC.2005.1470055
23. Somhom, S., Modares, A., Enkawa, T.: A self-organising model for the travelling salesman problem. J. Oper. Res. Soc. **48**(9), 919–928 (1997). https://doi.org/10.2307/3010491
24. Váňa, P., Faigl, J.: Optimal solution of the generalized Dubins interval problem finding the shortest curvature-constrained path through a set of regions. Auton. Robot. **44**(7), 1359–1376 (2020). https://doi.org/10.1007/s10514-020-09932-x

25. Yu, X., Hung, J.: A genetic algorithm for the Dubins traveling salesman problem. In: International Symposium on Industrial Electronics, pp. 1256–1261. IEEE (2012). https://doi.org/10.1109/ISIE.2012.6237270
26. Zhang, X., Chen, J., Xin, B., Peng, Z.: A memetic algorithm for path planning of curvature-constrained UAVs performing surveillance of multiple ground targets. Chin. J. Aeronaut. **27**(3), 622–633 (2014). https://doi.org/10.1016/j.cja.2014.04.024

Hyperbox-GLVQ Based on Min-Max-Neurons

Thomas Villmann(✉), T. Davies, and A. Engelsberger

Saxon Institute for Computational Intelligence and Machine Learning (SICIM), University of Applied Sciences Mittweida, Mittweida, Germany
villmann@hs-mittweida.de

Abstract. In this paper we propose the application of min-max-neurons for the use in generalized learning vector quantization (GLVQ) models, which correspond to min-max-prototypes. These prototypes can be identified with hyperboxes in the data space. Keeping the general GLVQ cost function, we redefine the Hebb-responsibilities for min-max-prototypes and derive consistent learning rules for stochastic gradient descent learning. We demonstrate that the resulting hyperbox-based GLVQ is capable to solve two illustrating toy classification tasks in robust manner, which can be dedicated to the use of robust min-max-prototypes. Finally, we give suggestions for future research for GLVQ based on min-max-prototypes.

Keywords: min-max-neurons · learning vector quantization

1 Introduction

Neural networks for supervised and unsupervised learning should be able to generalize after training. For this purpose, robustness regarding slightly varying input data is a highly welcome property [37]. However, this behavior is often difficult to achieve. In classification learning, one possibility is to built up large margin networks or models, which inherently optimize a classification margin. To the latter group belong support vector machines for binary classification problems [28,33] and generalized learning vector quantization [5,17,26]. Both methods can be taken as prototype based classifiers to a certain extend [2]. For deep learning several approaches are known, which enforce large margin decisions [19].

Data geometry related approaches for robust models like tangent learning approaches try to deal with variations in data using manifold learning methods [8,10,24]. The disadvantage of them is that the tangent learning is related to the local gradients of the cost function and, hence, are locally oriented and, therefore, not necessarily parallel to the data axes.

A simple alternative to these strategies are so-called min-max-neurons already proposed in the early 90-ies [31,32]. The response of these neurons to

T. Villmann et al. (Eds.): WSOM+2024, LNNS 1087, pp. 22–31, 2024.
https://doi.org/10.1007/978-3-031-67159-3_3

data are determined regarding fuzzy-memberships of the data to neuron receptive fields [6]. One of the most promising properties of min-max-neural-networks is the easy derivation of regarding decision rules [30]. Recently, a network built up from those neurons was proposed for classification learning, which can be trained by stochastic gradient descent learning (SGDL) [20]. In this approach, a cost function was derived such that minimization results in high classification accuracy based on the evaluation and adaptation of the min-max neurons. For this purpose, the hyperbox interpretation of the min-max neurons was adopted, which was already suggested in [31]. In consequence, the hyperboxes are in parallel to the feature axes of the data which allow the derivation of feature-based decision rules.

In this contribution, we propose to integrate min-max neurons to generalized learning vector quantizers (GLVQ) for robust classification learning. To do so, we introduce an appropriate prototype model and a corresponding dissimilarity measure to evaluate the response of a prototype to a given data sample. We show that the resulting scheme can be integrated into the SGDL-scheme of GLVQ. We illustrate the behavior of the resulting Hyperbox-GLVQ for two exemplary data sets.

2 Generalized Learning Vector Quantization Using Hyperbox-Prototypes

2.1 Min-Max Neurons

Min-max neurons were introduced in [31] by P. K. Simpson. Compared to standard perceptrons, min-max-neurons are equipped with two weight vectors $\mathbf{w}^l, \mathbf{w}^u \in \mathbb{R}^n$ with $w_j^l \leq w_j^u$. These are interpreted as lower and upper bound of a hyperbox $B\left(\mathbf{w}^l, \mathbf{w}^u\right)$ in $\mathbb{R}^n$. According to D.O. Hebb, the excitation for a neuron is assumed to reflect its match for a given stimulus [11]. For min-max-neurons it was declared to be the fuzzy-membership of stimulus $\mathbf{x} \in \mathbb{R}^n$ to the hyperbox $B\left(\mathbf{w}^l, \mathbf{w}^u\right)$. More precisely, using the quantities

$$m^u\left(\mathbf{x}, \mathbf{w}^u, \gamma, j\right) = \max\left(0, 1 - \max\left(0, \gamma \cdot \min\left(1, x_j - w_j^u\right)\right)\right)$$

and

$$m^l\left(\mathbf{x}, \mathbf{w}^l, \gamma, j\right) = \max\left(0, 1 - \max\left(0, \gamma \cdot \min\left(1, w_j^l - x_j\right)\right)\right)$$

Simpson defined the excitation/response

$$\hat{r}\left(\mathbf{x}, B\left(\mathbf{w}^l, \mathbf{w}^u\right), \gamma\right) = \frac{1}{n} \sum_{j=1}^{n} m^u\left(\mathbf{x}, \mathbf{w}^u, \gamma, j\right) + m^l\left(\mathbf{x}, \mathbf{w}^l, \gamma, j\right)$$

with $\gamma > 0$ is a sensitivity parameter that regulates the decrease of the membership function. Yet, according to a numerical analysis the behavior claimed in [31] can not be verified and should be replaced by

$$\hat{r}\left(\mathbf{x}, B\left(\mathbf{w}^l, \mathbf{w}^u\right), \gamma\right) = \frac{1}{n} \max_{j=1}^{n} m^u\left(\mathbf{x}, \mathbf{w}^u, \gamma, j\right) + m^l\left(\mathbf{x}, \mathbf{w}^l, \gamma, j\right) \qquad (1)$$

to achieve the desired result.[1] The final output of a min-max-neuron is obtained as

$$y = f\left(r\left(\mathbf{x}, B\left(\mathbf{w}^{l}, \mathbf{w}^{u}\right), \gamma\right) - \theta\right)$$

where $\theta \geq 0$ is the threshold and f is a monotonically increasing activation function as standard in perceptron learning [1,3,7,12]. Obviously, we have a good interpretation of a min-max-neuron: It is the more excited the better the hyperbox covers the stimulus.

A survey of min-max (or hyperbox) neural networks applications mainly in the context of fuzzy-learning can be found in [13]. Yet, a disadvantage of min-max neurons is the lack of differentiability, which make them not applicable in standard multi-layer perceptrons. Recently, in [20], this discrepancy was solved introducing smooth approximations of the min- and max-operators using parametrized sigmoid functions

$$\sigma_{\tau}\left(z\right) = \frac{1}{1 + \exp\left(\frac{z}{\tau}\right)} \tag{2}$$

with parameter $\tau > 0$. Further, the respective min-max-MLP is equipped with a final soft-max layer and the cost function is chosen as cross-entropy, as it is standard in (deep) MLP-learning for classification.

However, due to the (deep) MLP-architecture, the overall interpretability of min-max-MLP gets lost or is at least decreased compared to the simple interpretation of a singe min-max-neuron as it is frequently the case for (deep) MLP architectures [18,22]. Therefore, we will adopt min-max-neurons to be integrated in learning vector quantization classifiers, which are known to be interpretable and, additionally, show great robustness according to their inherent margin optimizer property [2].

2.2 Standard GLVQ

Learning vector quantization (LVQ) is a prototype based approach for classification learning, originally introduced by T. Kohonen in [14] to approximate Bayes classification decisions [15,16]. For given training data $\mathbf{x} \in \mathcal{X} \subseteq \mathbb{R}^n$ with class labels $c(\mathbf{x}) \in \mathcal{C} = \{1, \ldots, C\}$ a prototype network is designed using the set $\mathcal{W} = \{\mathbf{w}_1, \ldots, \mathbf{w}_M\} \subset \mathbb{R}^n$ of neuron weight vectors, which are denoted as prototype vectors in the vector quantization context. Thereby, each prototype is responsible for a class $c(\mathbf{w}_k) \in \mathcal{C}$ such that for each class $c \in \mathcal{C}$ at least one prototype is associated. The Hebb-response in LVQ is realized by

$$r\left(\mathbf{x}, \mathbf{w}_k, \gamma\right) = \exp\left(-\frac{d\left(\mathbf{x}, \mathbf{w}_k\right)}{\gamma}\right)$$

with parameter $\gamma > 0$ regulating the sensitivity in dependence on the dissimilarity value $d(\mathbf{x}, \mathbf{w}_k) \geq 0$. Frequently, the dissimilarity function is taken as the

[1] We thank one of the reviewers to this observation.

squared Euclidean distance. For a consideration regarding more general dissimilarities we refer to [21].

For a given prototype set $\mathcal{W}$, an unknown data sample $\mathbf{x} \in \mathbb{R}^n$ is classified according to the class $c(\mathbf{w}_s)$ where s is the most excited neuron (prototype) according to Hebb maximum excitation principle

$$s = \operatorname{argmax}_{k=1,\ldots M} \left(r\left(\mathbf{x}, \mathbf{w}_k, \gamma\right)\right) \tag{3}$$

known as Hebb-winner-takes-all (Hebb-WTA) rule. Training takes place by prototype adjustment to minimize misclassifications on the training set $\mathcal{X}$.

In [26] A. Sato & K. Yamada introduced for LVQ learning a smooth cost function $E_{GLVQ}(\mathcal{X}, \mathcal{W}) = \sum_{\mathbf{x} \in \mathcal{X}} E(\mathbf{x}, \mathcal{W})$ with local errors

$$E(\mathbf{x}, \mathcal{W}) = \mu \left(\frac{d^+(\mathbf{x}) - d^-(\mathbf{x})}{d^+(\mathbf{x}) + d^-(\mathbf{x})} \right) \tag{4}$$

approximating the overall classification error. Here, μ is a differentiable sigmoid function approximating the Heaviside function

$$H(z) = \begin{cases} 1 & z \geq 0 \\ 0 & z < 0 \end{cases} \tag{5}$$

and frequently chosen as the parametrized sigmoid $\sigma_T(z)$ from (2). Further, $d^+(\mathbf{x}) = d(\mathbf{x}, \mathbf{w}^+)$ is the dissimilarity of the, according the Hebb-response $r(\mathbf{x}, \mathbf{w}_k, \gamma)$, most excited prototype $\mathbf{w}^+$ among all prototypes associated to the class $c(\mathbf{x})$ of the training sample whereas $d^-(\mathbf{x}) = d(\mathbf{x}, \mathbf{w}^-)$ is the dissimilarity of the most excited prototype $\mathbf{w}^-$ of all incorrect prototypes.

The local errors $E(\mathbf{x}, \mathcal{W})$ from (4) allow stochastic gradient descent learning to adapt the prototype $\mathbf{w}^+$ and $\mathbf{w}^-$ in dependence on a given training sample $(\mathbf{x}, c(\mathbf{x}))$. It turns out that this scheme leads to a maximization of the class hypothesis margin if the dissimilarity d is induced by a semi-norm or semi-metric and, hence, ensuring robustness while keeping the model interpretability [5,23,36].

If the dissimilarity

$$d_{\Omega}(\mathbf{x}, \mathbf{w}_k) = \left(\boldsymbol{\Omega}(\mathbf{x} - \mathbf{w}_k)\right)^2 \tag{6}$$

with $\boldsymbol{\Omega} \in \mathbb{R}^{m \times n}$ and $m \leq n$ is used in GLVQ, also optimization of the matrix $\boldsymbol{\Omega}$ by SGDL is possible. This so-called relevance learning in GLVQ is known as matrix GLVQ (GMLVQ) for $m = n$ and limited rank matrix GLVQ (LiRaM-GLVQ) [4,27] for relevance learning originally proposed in [9].

2.3 Hyperbox-GLVQ

Following the idea of min-max-neurons, we associate to each prototypes $\mathbf{w}_k \in \mathbb{R}^n$ of a standard GLVQ a hyperbox $B_k \subset \mathbb{R}^n \times \mathbb{R}^n$ determined by the lower and upper bound vectors $\mathbf{w}_k^l = \mathbf{w}_k - \mathbf{b}_k$ and $\mathbf{w}_k^l = \mathbf{w}_k + \mathbf{b}_k$, respectively. Here, the bound vector $\mathbf{b}_k \in \mathbb{R}^n$ determines the range of the hyperbox for non-negative vector entries $b_{k,j} \geq 0$.[2] Those prototypes $\mathbf{W}_k = \{\mathbf{w}_k, \mathbf{b}_k\}$ we denote as min-max-prototypes or hyperbox-prototypes.

For the Hyperbox-GLVQ (H-GLVQ), we have to determine a respective Hebb-response. Following the min-max-neuron approach, a min-max-prototype $\mathbf{W}_k$ is excited regarding a data sample $\mathbf{x}$, if it is covered by the hyperbox $B_k = B\left(\mathbf{w}_k^l, \mathbf{w}_k^u\right)$, i.e. the indicator function

$$\mathbb{I}_{B_K}(\mathbf{x}) = \begin{cases} 1 & \mathbf{x} \in B_k \\ \\ 0 & \mathbf{x} \notin B_k \end{cases}$$

determines the excitation. To ensure differentiability we take a probabilistic smooth variant of $\mathbb{I}_{B_K}(\mathbf{x})$ in the following way: Similar to [20], we introduce the deviation vector

$$\boldsymbol{\Delta}_k = \mathbf{x} - \mathbf{w}_k \tag{7}$$

and estimate the probabilities

$$p_{k,j}(\mathbf{x}, \tau) = \sigma_\tau\left(\Delta_{k,j}^2 - b_{k,j}^2\right) \tag{8}$$

that a vector entry x_j violates the hyperbox bounds of B_k where we made use of the differentiable approximation σ_τ from (2) of the Heaviside function H from (5). Thus, the parameter $\tau > 0$ determines here the fuzziness of the hyperbox bounds. The resulting vector $\mathbf{p}_k(\mathbf{x}, \tau) \in [0,1]^n$ is used to define a dissimilarity measure

$$d_{\text{min-max}}(\mathbf{x}, \mathbf{W}_k, \tau) = n - \|\mathbf{p}_k(\mathbf{x}, \tau)\|_1$$

where $\|\cdot\|_1$ is the l_1-norm and, hence, is differentiable.[3] Hence, it can serve as a dissimilarity measure for the local error $E(\mathbf{x}, \mathcal{W})$ in GLVQ. Further, the Hebb response now is obtained as

$$r_{\text{min-max}}(\mathbf{x}, \mathbf{W}_k, \gamma, \tau) = \exp\left(-\frac{d_{\text{min-max}}(\mathbf{x}, \mathbf{W}_k, \tau)}{\gamma}\right) \tag{9}$$

required for the Hebb-WTA-rule (3) in H-GLVQ. Note, this Hebb-WTA is equivalent to the minimum dissimilarity WTA, which usually is applied in vector quantization.

Due to the differentiability of all quantities, SGDL learning for the min-max-prototypes $\mathbf{W}_k \in \mathcal{W}$ can be realized. Further, according to the hyperbox B_k interpretation, their data responsibilities can be easily recognized and corresponding decision rules are derived.

2 The non-negativeness $b_{k,j} \geq 0$ is only required for interpretation reasons. From mathematical as well as from algorithmic point of view, this constraint has not to be assumed.

3 Note, the quantity $\|\mathbf{p}_k(\mathbf{x}, \tau)\|_0$, with $\|\cdot\|_0$ being the nuclear norm, counts the boundary violations of $\mathbf{x}$ with respect to the hyperbox B_k.

3 Experimental Results

In this initial research we present results for two illustrating examples, to verify the standard behavior of the algorithmic approach. Both data sets are only two-dimensional such that we can achieve a visual impression of the results.

Spiral Data Set. The first data set is a two-dimensional toy data set of two twisted spirals. We can see in Fig. 1 that the min-max-prototypes try to cover the spirals in dependence on their class assignments. Obviously, one could easily derive from the H-GLVQ model a corresponding decision tree regarding the axes for class decision of the data samples which underlines the interpretability of the trained model.

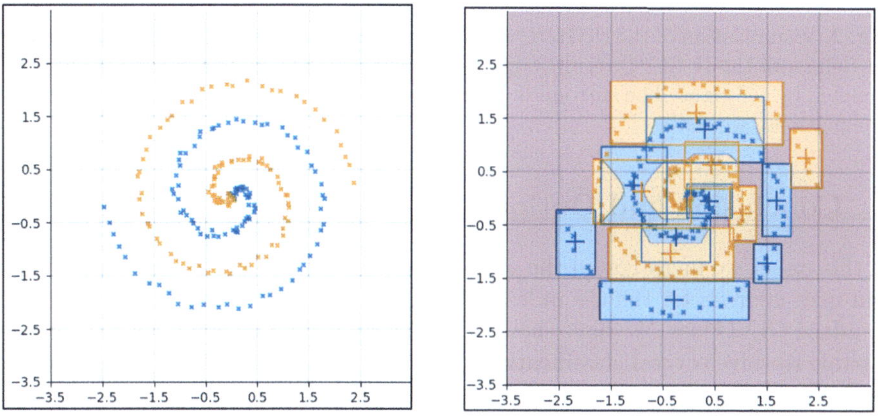

Fig. 1. Visualization of the H-GLVQ result for the two-dimensional spiral data set (left) and the H-GLVQ result (right). The centers of the min-max-prototypes are depicted by '+'-signs. The surrounding hyperboxes are marked accordingly.

The H-GLVQ was initialized with 20 min-max-prototypes for each class. During the learning process, if a hyperbox-size was squeezed to zero volume for min-max-prototype, this prototype was removed. Further, the parameter $\tau > 0$ for the sigmoid function σ_τ for the estimation of the hyperbox violation probabilities $p_{k,j}(\mathbf{x}, \tau)$ in (8) was initialized relatively large and then slowly decreased to approximate sharp hyperbox boundaries. After training, 14 min-max-prototypes survived (8/6 - for the classes) yielding an accuracy of 94.29%.

The second data set is the so-called moon-data set, as depicted in Fig. 2. The learning regime was similar to the spiral example. Here we started with 25 min-max-prototypes per class. Finally, we obtained 15 non-vanishing prototypes (12/3 - for the classes) resulting in an accuracy of 93.33%. For further explanations, see text.

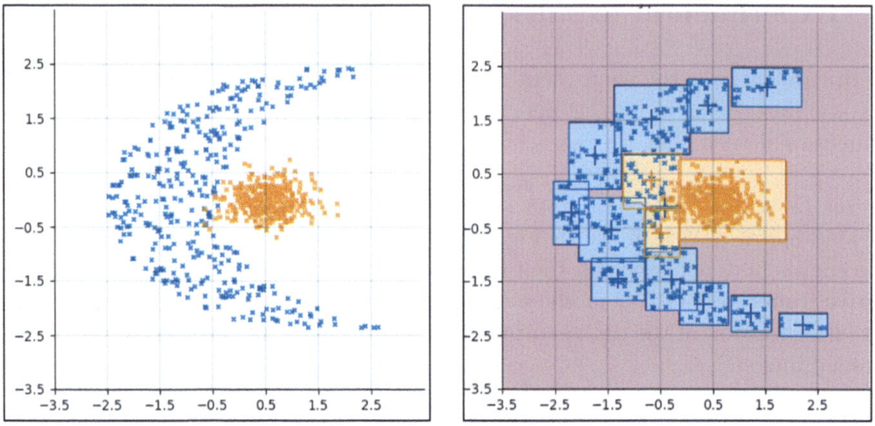

Fig. 2. Visualization of the H-GLVQ result for the two-dimensional artificial moon data set (left) and the H-GLVQ result (right). The centers of the min-max-protototypes are depicted by '+'-signs. The surrounding hyperboxes are marked accordingly. For further explanations, see text.

4 Conclusion and Outlook

In this paper we investigated the application of min-max-prototypes inspired by min-max-neurons for the use in learning vector quantization models keeping the standard GLVQ loss. We have shown that the resulting hyperbox-based GLVQ is capable to solve several classification tasks in robust manner. This behavior can be dedicated to the property that the min-max-prototypes represent hyperboxes, which should cover the given input. It was demonstrated that this approach can be seen as a specific Hebb excitation/response of the min-max-prototypes.

The resulting H-GLVQ is also somewhat similar to the restricted tangent GLVQ (rGTLVQ) [25,35]. The *difference* is that in rGTLVQ, the respective tangent hyperbox assigned to a prototype is in general not parallel to the coordinate axes but locally oriented with respect to the gradient of the loss function. Yet, taking the axes-parallel hyperboxes as prototypes allow a straightforward interpretation of the data response of the min-max-prototypes and the derivation of respective decision rules as suggested in [30].

Future work should investigate the performance improvement by metric learning according to (6) in GMLVQ. For this purpose, we could replace deviation vector $\boldsymbol{\Delta}_k$ from (7) by

$$\boldsymbol{\Delta}_k = \boldsymbol{\Omega}\left(\mathbf{x} - \mathbf{w}_k\right) \tag{10}$$

with $\boldsymbol{\Omega} \in \mathbb{R}^{m \times n}$ as known from GMLVQ/LiRaM-GLVQ. Generally, other dissimilarity measures according to the taxonomy proposed in [21] could be considered to evaluate point-to-set dissimilarities more properly. Further, the relation of H-GLVQ to probabilistic variants of GLVQ as proposed in [29,34] should be in the focus of ongoing considerations.

References

1. Biehl, M.: The Shallow and the Deep – A Biased Introduction to Neural Networks and old School Machine Learning. University of Groningen Press (2023)
2. Biehl, M., Hammer, B., Villmann, T.: Prototype-based models in machine learning. Wiley Interdisc. Rev.: Cogn. Sci. **7**(2), 92–111 (2016)
3. Bishop, C.M.: Pattern Recognition and Machine Learning. Springer, Heidelberg (2006)
4. Bunte, K., Schneider, P., Hammer, B., Schleif, F.-M., Villmann, T., Biehl, M.: Limited rank matrix learning, discriminative dimension reduction and visualization. Neural Netw. **26**(1), 159–173 (2012)
5. Crammer, K., Gilad-Bachrach, R., Navot, A., Tishby, N.: Margin analysis of the LVQ algorithm. In: Becker, S., Thrun, S., Obermayer, K. (eds.) Advances in Neural Information Processing (Proceedings of the NIPS 2002), vol. 15, pp. 462–469. MIT Press, Cambridge (2003)
6. Gabrys, B., Bargiela, A.: General fuzzy min-max neural network for clustering and classification. IEEE Trans. Neural Netw. **11**(3), 769–783 (2000)
7. Goodfellow, I., Bengio, Y., Courville, A.: Deep Learning. MIT Press, Cambridge (2016)
8. Haasdonk, B., Keysers, D.: Tangent distance kernels for support vector machines. In: Kasturi, R., Laurendeau, D., Suen, C. (eds.) Proceedings of the 16th International Conference on Pattern Recognition (ICPR), Québec City, vol. 2, pp. 864–868. IEEE Press, Los Alamitos (2002)
9. Hammer, B., Villmann, T.: Generalized relevance learning vector quantization. Neural Netw. **15**(8–9), 1059–1068 (2002)
10. Hastie, T., Simard, P., Säckinger, E.: Learning prototype models for tangent distance. In: Tesauro, G., Touretzky, D.S., Leen, T.K. (eds.) Advances in Neural Information Processing Systems, vol. 7, pp. 999–1006. MIT Press (1995)
11. Hebb, D.O.: The Organization of Behavior. A Neuropsychological Theory. Wiley, New York (1949)
12. Hertz, J.A., Krogh, A., Palmer, R.G.: Introduction to the Theory of Neural Computation. Santa Fe Institute Studies in the Sciences of Complexity: Lecture Notes, vol. 1. Addison-Wesley, Redwood City (1991)
13. Khuat, T., Ruta, D., Gabrys, B.: Hyperbox-based machine learning algorithms: a comprehensive survey. Soft. Comput. **25**, 1325–1363 (2021)
14. Kohonen, T.: Learning vector quantization. Neural Netw. **1**(Supplement 1), 303 (1988)
15. Kohonen, T.: Improved versions of learning vector quantization. In: Proceedings of the IJCNN-90, International Joint Conference on Neural Networks, San Diego, vol. I, pp. 545–550. IEEE Service Center, Piscataway (1990)
16. Kohonen, T.: Statistical pattern recognition revisited. In: Eckmiller, R. (ed.) Advanced Neural Networks, pp. 137–144. Elsevier, Amsterdam (1990)
17. Kusunoki, Y., Wakou, C., Tatsumi, K.: Maximum-margin model for nearest prototype classifiers. J. Adv. Comput. Intell. Intell. Inform. **22**(4) (2018)
18. Lisboa, P., Saralajew, S., Vellido, A., Fernández-Domenech, R., Villmann, T.: The coming of age of interpretable and explainable machine learning models. Neurocomputing **535**, 25–39 (2023)
19. Matyasko, A., Chau, L.P.: Margin maximization for robust classification using deep learning. In: Proceedings of the International Joint Conference on Neural Networks (IJCNN), pp. 300–307. IEEE Press (2017)

20. Mayr Lima Martins, D., Lülf, C., Gieseke, F.: End-to-end neural network training for hyperbox-based classification. In: Verleysen, M. (ed.) Proceedings of the 31st European Symposium on Artificial Neural Networks, Computational Intelligence and Machine Learning (ESANN 2023), Bruges (Belgium), Louvain-La-Neuve, Belgium, pp. 193–198 (2023). i6doc.com
21. Nebel, D., Kaden, M., Villmann, A., Villmann, T.: Types of (dis−)similarities and adaptive mixtures thereof for improved classification learning. Neurocomputing **268**, 42–54 (2017)
22. Samek, W., Monatvon, G., Vedaldi, A., Hansen, L.K., Müller, K.-R. (eds.): Explainable AI: Interpreting, Explaining and Visualizing Deep Learning. LNAI, vol. 11700. Springer, Heidelberg (2019). https://doi.org/10.1007/978-3-030-28954-6
23. Saralajew, S., Holdijk, L., Villmann, T.: Fast adversarial robustness certification of nearest prototype classifiers for arbitrary seminorms. In: Larochelle, H., Ranzato, M., Hadsell, R., Balcan, M.F., Lin, H. (eds.) Proceedings of the 34th Conference on Neural Information Processing Systems (NeurIPS 2020), vol. 33, pp. 3635–13650. Curran Associates, Inc. (2020)
24. Saralajew, S., Villmann, T.: Adaptive tangent metrics in generalized learning vector quantization for transformation and distortion invariant classification learning. In: Proceedings of the International Joint Conference on Neural networks (IJCNN), Vancouver, pp. 2672–2679. IEEE Computer Society Press (2016)
25. Saralajew, S., Villmann, T.: Restricted tangent metrics for local data dissimilarities - mathematical treatment of the corresponding constrained optimization problem. Mach. Learn. Rep. **11**(MLR-01-2017) (2017, submitted). http://www.techfak.uni-bielefeld.de/~fschleif/mlr/mlr_01_2017.pdf. ISSN 1865-3960
26. Sato, A., Yamada, K.: Generalized learning vector quantization. In: Touretzky, D.S., Mozer, M.C., Hasselmo, M.E. (eds.) Advances in Neural Information Processing Systems. Proceedings of the 1995 Conference, vol. 8, pp. 423–429. MIT Press, Cambridge (1996)
27. Schneider, P., Hammer, B., Biehl, M.: Adaptive relevance matrices in learning vector quantization. Neural Comput. **21**, 3532–3561 (2009)
28. Schölkopf, B., Smola, A.: Learning with Kernels. MIT Press, Cambridge (2002)
29. Seo, S., Obermayer, K.: Soft learning vector quantization. Neural Comput. **15**, 1589–1604 (2003)
30. Shinde, S.V., Kulkarni, U.V., Chaudary, A.N.: Extracting the classification rules from general fuzzy min-max neural network. Int. J. Comput. Appl. **121**(23), 1–7 (2015)
31. Simpson, P.K.: Fuzzy min-max neural networks - part 1: classsification. IEEE Trans. Neural Netw. **3**(5), 776–786 (1992)
32. Simpson, P.K.: Fuzzy min-max neural networks - part 2: clustering. IEEE Trans. Neural Netw. **1**(1), 32–45 (1993)
33. Steinwart, I., Christmann, A.: Support Vector Machines. Information Science and Statistics. Springer, Heidelberg (2008). https://doi.org/10.1007/978-0-387-77242-4
34. Villmann, A., Kaden, M., Saralajew, S., Villmann, T.: Probabilistic learning vector quantization with cross-entropy for probabilistic class assignments in classification learning. In: Rutkowski, L., Scherer, R., Korytkowski, M., Pedrycz, W., Tadeusiewicz, R., Zurada, J.M. (eds.) ICAISC 2018. LNCS (LNAI), vol. 10841, pp. 724–735. Springer, Cham (2018). https://doi.org/10.1007/978-3-319-91253-0_67

35. Villmann, T., Saralajew, S.: Restricted tangent distances in learning vector quantization. Mach. Learn. Rep. **11**(MLR-03-2017), 5–6 (2017). http://www.techfak.uni-bielefeld.de/~fschleif/mlr/mlr_03_2017.pdf. ISSN 1865-3960
36. Voráček, V., Hein, M.: Provably adversarially robust nearest prototype classifiers. In: Proceedings of the 39th International Conference on Machine Learning (ICML). Proceeding of Machine Learning Research (PMLR), vol. 162, pp. 22361–22383 (2022)
37. Xu, H., Mannor, S.: Robustness and generalization. Mach. Learn. **86**, 391–423 (2012)

Sparse Clustering with K-Means - Which Penalties and for Which Data?

Marie Chavent[1], Marie Cottrell[2], Alex Mourer[2], and Madalina Olteanu[3](✉)

[1] Univ. Bordeaux, CNRS, INRIA, Bordeaux INP, IMB, UMR 5251, Talence, France
marie.chavent@math.u-bordeaux.fr

[2] SAMM, EA 4543 - Université Paris 1 Panthéon-Sorbonne, Paris, France
{marie.cottrell,alex.mourer}@univ-paris1.fr

[3] CEREMADE UMR 7534, Paris Dauphine University PSL, Paris, France
olteanu@ceremade.dauphine.fr

Abstract. While high dimensionality and the selection of meaningful features is usually a burden in machine learning, it is even more so in the case of unsupervised learning and particularly in clustering. The presence of uninformative features may bias significantly the results of distance-based methods such as K-means for instance. For tackling this issue, different versions of sparse K-means have been introduced, building on the idea of adding some penalty terms in the loss function and resulting into automatic feature selection and/or weighting. This paper investigates the connections between some of these methods, and particularly the differences induced by the choices of the penalty terms. It also illustrates the algorithms and model selection tools made available through a recently implemented R package, vimpclust.

Keywords: Sparse clustering · K-means · Mixed data · R package

1 Introduction

In the current context of massive data collection, clustering high-dimensional data is an extremely challenging topic. Indeed, clustering is already a difficult problem in practice, since the structure one is looking for is not priorly known, and different levels of structure may be revealed depending on the granularity of the data and on the number of clusters considered. When the number of features is high, some of them may be purely noise and uninformative for the clustering structure, others, possibly the same, could be correlated and induce a bias consisting in considering them more important. Selecting the features carrying the meaningful information is thus a crucial step for obtaining good performances in clustering tasks. One important question arising is whether one should perform feature selection before clustering, or couple the two tasks together. The present manuscript focuses on the latter. More particularly, we will consider here generalisations of the K-means algorithm for high-dimensional data, which aim at both clustering and retrieving a sparse representation of the

T. Villmann et al. (Eds.): WSOM+2024, LNNS 1087, pp. 32–41, 2024.
https://doi.org/10.1007/978-3-031-67159-3_4

clusters. K-means may be seen as both a Gaussian mixture model with some specific constraints on the covariance matrix, and a distance-based heuristic searching for the partition maximizing the between sum-of-squares. With the latter point of view, it is obvious that high-dimensionality and its related issues such as distance concentration, empty space etc. will impact the performances of the algorithm [1,9]. With this in mind, which are the options for dealing with the issue of high-dimensionality in K-means training?

Several approaches have been proposed in the literature, and are based on introducing penalties or regularization terms which allow for sparsity. Some of these approaches consist in introducing a set of weights which are applied to the features, and which are subject to some regularization. Among these, some build on combinations of ℓ_1-ℓ_2 penalties [3,5,7,10], others on combinations of ℓ_0-ℓ_∞ [2], while some others take into account the correlation structure of the features [6]. Other approaches introduce penalty terms directly on the average of the cluster (the data is priorly centered and scaled) [8]. Most of these methods deal exclusively with numerical features and do not discuss the case of mixed data, containing both categorical and numerical features.

The present manuscript focuses on a particular sparse generalisation of the K-means algorithm. First, a set of weights associated to the features is introduced, and the weighted version of the between sum-of-squares is considered as the criterion to optimize. Feature selection is achieved by applying ℓ_1 and ℓ_2 constraints on the weights. At least two different writings of the problem and two associated algorithms were proposed in the literature. The first one was introduced in [10], where the ℓ_1 penalty is a supplementary constraint, while a second one was described in [3], where the ℓ_1 penalty is directly introduced in the criterion to optimize. Although extremely similar, the two writings of the problem lead to quite different solutions, actually. The differences between them in terms of performances, both clustering and feature selection, will be investigated in the following. After recalling both algorithms in Sect. 2, a simulation study is carried out in Sect. 3 in order to illustrate the differences between the two. Specifically, the tuning of the regularization parameter and its impact on the performances will be investigated. Eventually, Sect. 4 will present the main facilities of an R package, `vimpclust`, which implements some of the methods discussed in this paper.

2 Sparse K-Means

Before describing the two procedures, let us introduce the necessary notations. The data, which is supposed to be made of numerical features only, is represented by $\mathbf{X} \in \mathbb{R}^{n \times p}$. We thus consider n input data and p features, with p possibly very large. $\mathbf{x}_i$ is the i-th input valued in $\mathbb{R}^p$, while $\mathbf{x}^j$ is the j-th feature in $\mathbb{R}^n$. The data is supposed to be priorly scaled to zero mean and unit variance. One will also consider a partition $\mathcal{C} = \{C_1, ..., C_K\} \in \mathcal{P}_K$, where K is the number of clusters, and $\mathcal{P}_K$ is the set of possible partitions into K clusters. In the following, K will be considered as priorly known and fixed throughout. How to select K is out of the scope of the present paper, in order to remove a complexity layer.

Sparsity will be dealt with by introducing a set of weights associated to the features and by applying some regularization constraints on them. Uninformative features should be removed when their weights are set to 0. The weights are given by a vector $\mathbf{w} \in \mathbb{R}^p$, and one may write a weighted version of the between sum-of-squares (BSS), or, alternatively, of the within sum-of-squares (WSS):

$$BSS(\mathcal{C}, \mathbf{w}) = \mathbf{w}^\top \mathbf{b} = \sum_{j=1}^{p} w_j b_j = \sum_{j=1}^{p} w_j \sum_{k=1}^{K} n_k (\overline{\mathbf{x}}_k^j - \overline{\mathbf{x}}^j)^2, \tag{1}$$

where b_j is the BSS associated to feature j, n_k is the size of cluster k, $\overline{\mathbf{x}}_k^j$ is the average of feature j in cluster k, and $\overline{\mathbf{x}}^j$ is the overall average of feature j. One should remark that in the weighted writing, minimizing the weighted WSS or maximizing the weighted BSS are no longer equivalent.

2.1 First Formulation of the Problem

From a chronological perspective, sparse K-means was introduced in [10] with the following writing:

$$\underset{\mathcal{C}, \mathbf{w}}{\text{maximize}} \quad \mathbf{w}^\top \mathbf{b} \quad \text{subject to} \quad \|\mathbf{w}\|_1 \leq s, \|\mathbf{w}\|_2 \leq 1, w_j \geq 0,\ j = 1, \ldots, p, \tag{2}$$

where $\mathbf{b} \in \mathbb{R}^p$ is the vector containing the BSS per feature for the partition $\mathcal{C}$, and $s \geq 1$ is the tuning parameter. Solving problem (2) is achieved through an iterative procedure, which alternates between keeping one of $\mathcal{C}$ and $\mathbf{w}$ fixed, and optimizing with respect to the other. In practice, this amounts to the following procedure:

1. Initialize the feature weights $w_1 = \ldots = w_p = \frac{1}{\sqrt{p}}$ (no preferred feature).
2. Repeat until convergence:
   (a) For fixed $\mathbf{w}$, find the partition $\mathcal{C}$ by training the usual K-means algorithm on the data scaled as $\tilde{\mathbf{x}}^j = \sqrt{w_j}\mathbf{x}^j, \forall j = 1, ..., p$.
   (b) For fixed $\mathcal{C}$, find the vector of weights $\mathbf{w}$ by solving the optimization problem above.

Several remarks are to be made with the respect to the proposed algorithm. First, one should notice that when keeping $\mathbf{w}$ fixed, the optimization problem indeed reduces to finding the best partition $\mathcal{C}$ which maximizes the BSS criterion, where each feature $\mathbf{x}^j$ has been scaled with respect to the weight w_j. Second, when $\mathcal{C}$ is fixed, one may find an analytical solution to problem 2 using the soft-thresholding operator:

$$\widehat{\mathbf{w}} = \frac{\mathbf{S}(\mathbf{b}, \Delta)}{\|\mathbf{S}(\mathbf{b}, \Delta)\|_2}, \text{ where } \mathbf{S}(\mathbf{b}, \Delta) = (S(b_1, \Delta), \ldots, (S(b_p, \Delta))^\top, \tag{3}$$

and

$$S(b_j, \Delta) = \frac{b_j}{|b_j|}(|b_j| - \Delta)_+ = (b_j - \Delta)_+ \tag{4}$$

The soft-thresholding operator sets to zero all feature weights which b_j is smaller than Δ, and shrinks the other towards zero by the same quantity. One may further remark here that the threshold Δ depends on the position of s relative to $\frac{\|\mathbf{b}\|_1}{\|\mathbf{b}\|_2}$:

$$\begin{cases} \Delta = 0 \text{ if } \frac{\|\mathbf{b}\|_1}{\|\mathbf{b}\|_2} \leq s, \\ \Delta \text{ is such that } \frac{\|\mathbf{S}(\mathbf{b},\Delta)\|_1}{\|\mathbf{S}(\mathbf{b},\Delta)\|_2} = s \text{ if } \frac{\|\mathbf{b}\|_1}{\|\mathbf{b}\|_2} > s. \end{cases} \tag{5}$$

In other words, either the solution $\widehat{\mathbf{w}} = \frac{\mathbf{b}}{\|\mathbf{b}\|_2}$ satisfies the ℓ_1-norm constraint $(\frac{\|\mathbf{b}\|_1}{\|\mathbf{b}\|_2} \leq s)$ and then $\Delta = 0$, or Δ is used for soft-thresholding $\mathbf{b}$ such that $\widehat{\mathbf{w}} = \frac{\mathbf{S}(\mathbf{b},\Delta)}{\|\mathbf{S}(\mathbf{b},\Delta)\|_2}$ satisfies the constraint $(\frac{\|\mathbf{S}(\mathbf{b},\Delta)\|_1}{\|\mathbf{S}(\mathbf{b},\Delta)\|_2} = s)$. A binary search procedure can then be used to find Δ according to s, according to [10]. Since problem (2) is non-convex, the iterative strategy will generally not achieve a global optimum. It is however guaranteed that each iteration will improve the cost function.

2.2 Second Formulation of the Problem

A different writing of the sparse K-means with a ℓ_1-penalty was proposed in [3]. Instead of the ℓ_1 constraint, a penalty term is directly added to the cost function, and the optimization problem becomes:

$$\underset{\mathcal{C},\mathbf{w}}{\text{maximize}} \quad \mathbf{w}^\top \mathbf{b} - \lambda\|\mathbf{w}\|_1 \quad \text{subject to} \quad \|\mathbf{w}\|_2 \leq 1, w_j \geq 0,\ j = 1, \ldots, p, \tag{6}$$

where $\mathbf{b} \in \mathbb{R}^p$ is the vector containing the BSS per feature for the partition $\mathcal{C}$, and $\lambda \geq 0$ is the tuning parameter. The optimization procedure is similar to the one in the previous section: one starts with uniform weights, and then alternates between solving the problem in $\mathcal{C}$ for fixed $\mathbf{w}$, and solving the problem in $\mathbf{w}$ for fixed $\mathcal{C}$. When $\mathbf{w}$ is fixed, one gets exactly the same step where she must train a K-means on the original features scaled with respect to the weights. When $\mathcal{C}$ is fixed, the updates of the weights may be written again using the soft-thresholding operator:

$$\widehat{\mathbf{w}} = \begin{cases} \frac{\mathbf{S}(\mathbf{b},\lambda)}{\|\mathbf{S}(\mathbf{b},\lambda)\|_2} & \text{if } \mathbf{S}(\mathbf{b},\lambda) \neq 0, \\ 0 & \text{if } \mathbf{S}(\mathbf{b},\lambda) = 0. \end{cases} \tag{7}$$

As above, the iterative strategy will not achieve a global optimum, but is guaranteed to improve the cost function at each step.

2.3 Discussion and Hyperparameter Tuning

Although the two algorithms look very similar, they do not actually lead to the same solutions. Indeed, while λ is fixed throughout the whole training procedure and represents the quantity by which shrinking the current values of the BSS per feature, Δ will change at each new iteration, depending on s and the current partition $\mathcal{C}$ (a link between Δ and λ may be made only when the partition $\mathcal{C}$ is fixed). While the first algorithm aims at preserving the most significant features up to some total level of "information" s, the second will discard all features which BSS does not achieve at least a level of information λ per feature. With the above, one may easily remark that the first algorithm will generally preserve at least some features, while the second may be quite conservative and discard all of them.

A natural question arising then is how to select an appropriate value for tuning the hyper-parameters s and λ, and whether there are some bounds for the values among which to look for. The reader should recall here that all features are supposed to be scaled to zero mean and unit variance. Although this is not systematically done in practice, it is a crucial step both for avoiding biases in favour of features bearing more heterogeneity, and for tuning the penalty terms. In practice, for selecting an appropriate value for s or λ, one will consider a grid of possible values, and evaluate some criterion (BSS, WSS, weighted versions of the previous, gap statistic, ...) through its regularization path. In [10], the tuning parameter s is related to $\frac{\|\mathbf{b}\|_1}{\|\mathbf{b}\|_2}$, where $\mathbf{b}$ is the BSS vector. Because of the norm inequality $\|\mathbf{b}\|_2 \leq \|\mathbf{b}\|_1 \leq \sqrt{p}\,\|\mathbf{b}\|_2$, one may remark that if $s \geq \sqrt{p}$, then $\frac{\|\mathbf{b}\|_1}{\|\mathbf{b}\|_2} \leq s$ for all partitions ($\forall \mathbf{b}$) and then $\Delta = 0$ at each iteration. All the features will be selected if $s \geq \sqrt{p}$. On the other hand, if $s < 1$, then $\frac{\|\mathbf{b}\|_1}{\|\mathbf{b}\|_2} > s$ for all partitions, and Δ is chosen at each iteration so that $\frac{\|\mathbf{S}(\mathbf{b},\Delta)\|_1}{\|\mathbf{S}(\mathbf{b},\Delta)\|_2} = s$, which is impossible. Then, a reasonable choice for a grid search will be to consider $s \in [1, \sqrt{p}]$. In [3], the tuning parameter λ is exactly the threshold used for updating feature weights through soft-thresholding. Since the features are scaled to unit variance, an immediate choice would be $\lambda \in [0, 1]$.

3 Numerical Illustrations

Since the two algorithms are not equivalent, as neither are the roles of the tuning parameters and the search grids for the most appropriate values, this section aims at comparing them, both in terms of clustering abilities and feature selection. For this purpose, two synthetic data generation scenarios were used.

3.1 Synthetic Data - Scenario A

In this example, the data is simulated according to a mixture of Gaussian distributions, with means $\mu_1 = (-5, 0)$, $\mu_2 = (0, 0)$, $\mu_3 = (5, 0)$, $\mu_4 = (0, -5)$, $\mu_5 = (0, 5)$. The covariance matrices are diagonal, with the diagonal equal to 1. Uninformative features are also added, through standard Gaussian distributions. The number of uninformative features varies in $\{1, 5\}$, so that the final number of features is $p = 3$ or 7. All features are mutually independent. Within each of the five clusters k, $n_k = 100$ input values were simulated. Eventually, the resulting data was scaled to zero mean and unit variance, in order to give prior equal importance to all features. For $p = 3$, the data is illustrated in Fig. 1.

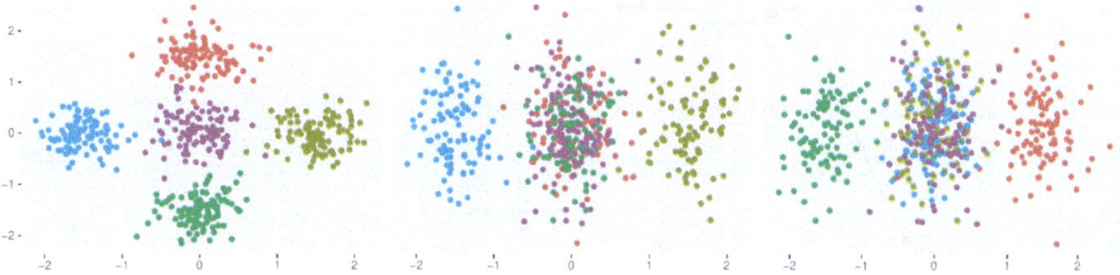

Fig. 1. Data simulated with scenario A. Left: features 1 and 2. Middle: features 1 and 3. Right: features 2 and 3. Only the first two features are informative.

Focusing on One Data Set. The two algorithms were trained on the data in Fig. 1. The tuning parameters s and λ were selected using the gap statistic as defined in [10] ($B = 10$ permutations), and a regular grid with 20 values in $[1, \sqrt{p}]$ for s, and in $[0, 1]$ for λ. The resulting regularization paths in terms of gap statistic, feature weights, BSS and weighted BSS are represented in Fig. 2 for the first algorithm, and Fig. 3 for the second. The values for tuning parameter (s or λ) maximizing the gap statistic are drawn with a vertical red line. First, one should notice that the regularization paths should be read the opposite way: smaller values of s and larger values of λ are equivalent to large penalty terms. Because of the constraints on s, one is guaranteed to select at least one feature, whatever the value of the tuning parameter. For λ, large values may lead to discarding all features, and to obtaining a degenerate result (these values were not plotted on the graphs). Second, one may remark that the best tuning parameter according to the gap statistic is not necessarily the same according to the criterion optimized by the sparse K-means algorithms (weighted BSS). Nevertheless, whether one decides to use the gap statistic or the value of the tuning parameter achieving a good trade-off between the weighted BSS and the number of selected features, the features eventually selected and the clustering performances are very similar (identical for this toy example).

Performances. In order to make the comparison more general, $R = 25$ different data sets were simulated according to Scenario A. For each of them, the best tuning parameters s and λ were selected using a regular grid with 20 values in $[1, \sqrt{p}]$ for s, and in $[0, 1]$ for λ. The criterion for selecting the tuning parameter was the gap statistic ($B = 10$ permutations). Once the best tuning parameter was selected, the two algorithms were trained for this particular value. For each of the resulting clusterings, we computed the adjusted Rand index (ARI) for assessing clustering, the ratio of true features detected or the true positive rate (TPR) and the ratio of false features detected or the false positive rate (FPR) for assessing feature selection. The results are summarized in Table 1 (mean and standard deviation). The same random generator seeds were used to control for initialization. The results in Table 1 provide clear insights on the performances of the two approaches. In terms of clustering performances, both are very similar, particularly when the number of uninformative features is equal to 1. The main

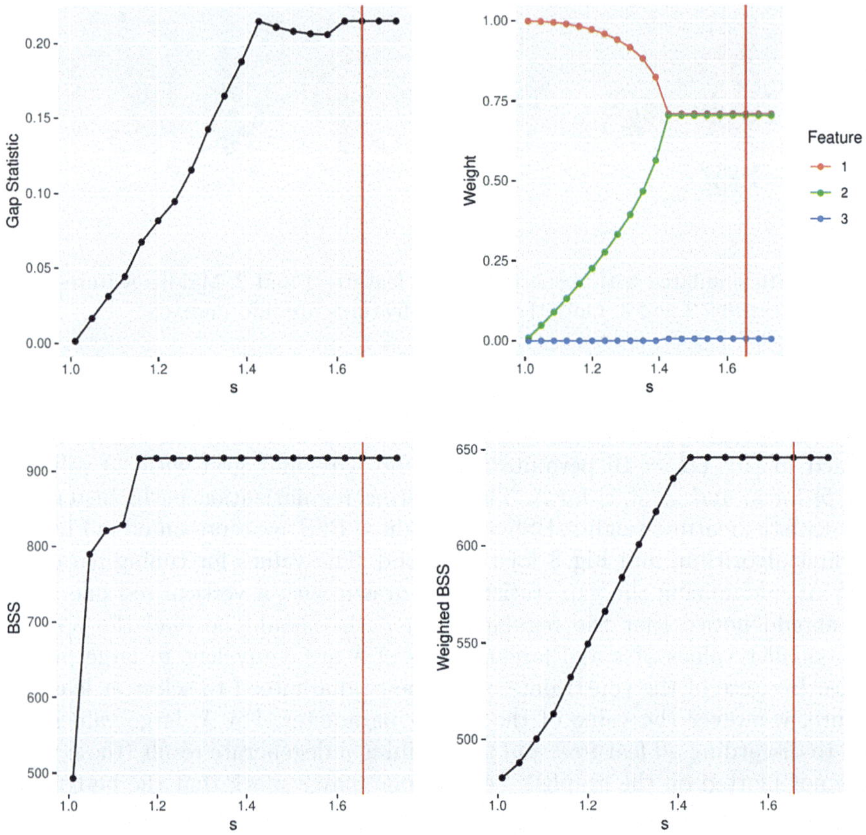

Fig. 2. Regularization paths - first algorithm. The vertical red line corresponds to the optimal tuning parameter according to the gap statistic.

Table 1. Aggregated results (mean and sd for 25 replications). Scenario A.

	Adjusted Rand Index		True Positive Rate		False Positive Rate	
	$p = 3$	$p = 7$	$p = 3$	$p = 7$	$p = 3$	$p = 7$
Algorithm 1	0.97 (0.01)	0.92 (0.19)	100. (0)	100.00 (0)	100.00 (0)	88.80 (31.13)
Algorithm 2	0.97 (0.01)	0.97 (0.01)	100.00 (0)	100.00 (0)	0.00 (0)	0.00 (0)

differences arise however in terms of feature selection. The informative features for the clustering structure are correctly selected by both approaches. The uninformative features are "seen" only by the second approach. One may remark that this behaviour may come from the fact that the gap statistic may not be the best criterion to tune the hyperparameter, and this issue may be further investigated in the future.

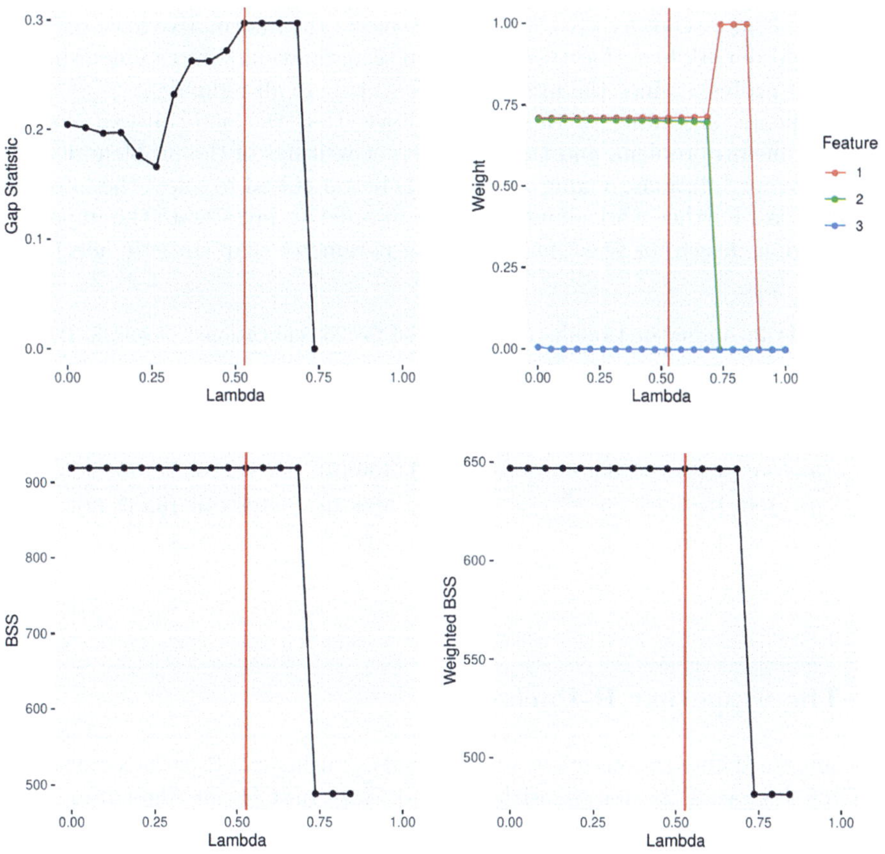

Fig. 3. Regularization paths - second algorithm. The vertical red line corresponds to the optimal tuning parameter according to the gap statistic.

3.2 Synthetic Data - Scenario B

The second scenario was inspired by one of the examples in [10]. It is based on a mixture of Gaussian distributions with three components in dimension 1, $\mu_1 = -3$, $\mu_2 = 0$, $\mu_3 = 3$. The first 50 features in the dataset are all simulated according to the same mixture, and independently. Within each of the three clusters k, $n_k = 100$ input values were simulated. Noise was added using standard Gaussian distributions, with a number of noise features varying in $\{0, 50, 100\}$, which amounts to $p \in \{50, 100, 150\}$. Eventually, the resulting data was scaled to zero mean and unit variance. The above procedure was used for simulating several datasets and comparing the performances. The results are presented in Table 2. They are very similar to the previous ones and confirm the same behaviour: both algorithms have very similar performances in terms of clustering, but the second one is better for feature selection. While the first algorithm always selects all the features, including the uninformative ones, the second algorithm is more conservative, and mostly selects the informative features only. One should remark here that this simulation scenario results into a much simpler clustering problem, since the informative features are all redundant.

It appears, according to the results above, that the non-convexity of the sparse K-means problem, and the two different writings of the problem, lead to quite different solutions, among which one is better suited to select the informative features. Further work should be done in order to investigate the impact of the criterion chosen for selecting the tuning parameter (gap statistic here).

Table 2. Aggregated results (mean and sd for 25 replications). Scenario B.

	Adjusted Rand Index		True Positive Rate		False Positive Rate	
	Algo 1	Algo 2	Algo 1	Algo 2	Algo 1	Algo 2
$p = 50$	1 (0)	1 (0)	100 (0)	100 (0)		
$p = 100$	1 (0)	1 (0)	100 (0)	100 (0)	100 (0)	16 (37.42)
$p = 150$	1 (0)	1 (0)	100 (0)	100 (0)	100 (0)	8 (27.69)

4 The `vimpclust` R-Package

The numerical illustrations above were carried out using two R packages available on CRAN: `sparcl` [11] for Algorithm 1, and `vimpclust` [4] for Algorithm 2. We would like to insist here on the importance of efficient and well-documented implementations, which are so much helpful for users and practitioners, as well as for running straightforward comparisons of methods. `sparcl` contains sparse versions of K-means and hierarchical clustering for numerical features. The K-means problem is solved with Algorithm 1. The package contains one function allowing to compute the gap statistic for a range of possible values for the tuning parameter. Additional functionalities such as illustrations of the regularizations paths are however quite limited. `vimpclust` implements sparse K-means using the second algorithm. It does not compute the gap statistic, but it contains plot functions of several regularization paths (BSS, weighted BSS, number of selected features, weights, ...). The user may inspect these curves and decide on the best value for the tuning parameter, by looking for a trade-off between the clustering performance and the number of selected features.

An important aspect of `vimpclust` is the generalisation of sparse K-means to more complex data. Indeed, the package allows training the algorithm for groups of numerical features, mixed data (categorical and numerical features) pre-processed appropriately, and groups of mixed features. This is achieved by replacing the ℓ_1 penalty term by a ℓ_1-group penalty term, or by a mixture of ℓ_1 and ℓ_1-group penalties. The resulting algorithms for groups of numerical features and for mixed features are detailed for instance in [3].

5 Conclusion and Perspectives

As mentioned in the introduction, clustering high-dimensional data is extremely challenging, since possibly many features may be uninformative, and the noise they carry may mask the actual structure. We focused here on sparse clustering achieved through a ℓ_1 penalty. Since the problem is non-convex, different writings of it lead to different heuristics, which may have quite different outcomes. The performances of two algorithms were studied, and it was shown that one of them was better for feature selection task. Nevertheless, there are several points that should be considered with more details in the future. One of them is the criterion used for selecting the best value for the tuning parameter. While the gap statistic appears as a natural choice, its impact should be studied and compared with other possibilities. Another issue to consider is the grid of values for the tuning parameters. Regular grids were used here, but other options are to be studied. Eventually, we should mention here that other approaches for sparse K-means use ℓ_1 penalties applied directly to the vector of means [8]. A more thorough comparison including these is considered for the future.

References

1. Bellman, R.: Adaptive Control Processes: A Guided Tour. Princeton University Press, Princeton (1961)
2. Chang, X., Wang, Y., Li, R., Xu, Z.: Sparse k-means with $l_\infty - l_0$ penalty for high-dimensional data clustering. Stat. Sin. **28**(3), 1265–1284 (2018)
3. Chavent, M., Lacaille, J., Mourer, A., Olteanu, M.: Sparse k-means for mixed data via group-sparse clustering. In: ESANN 2020-28th European Symposium on Artificial Neural Networks, Computational Intelligence and Machine Learning, vol. 978 (2020)
4. Chavent, M., Mourer, A., Olteanu, M.: vimpclust: variable importance in clustering (2021). https://cran.r-project.org/web/packages/vimpclust/
5. Diallo, A.W., Niang, N., Ouattara, M.: Sparse subspace k-means. In: 2021 International Conference on Data Mining Workshops (ICDMW), pp. 678–685. IEEE (2021)
6. Gong, W., Zhao, R., Grünewald, S.: Structured sparse k-means clustering via Laplacian smoothing. Pattern Recogn. Lett. **112**, 63–69 (2018)
7. Kondo, Y., Salibian-Barrera, M., Zamar, R.: RSKC: an R package for a robust and sparse k-means clustering algorithm. J. Stat. Softw. **72**, 1–26 (2016)
8. Raymaekers, J., Zamar, R.H.: Regularized k-means through hard-thresholding. J. Mach. Learn. Res. **23**(93), 1–48 (2022)
9. Verleysen, M., François, D.: The curse of dimensionality in data mining and time series prediction. In: Cabestany, J., Prieto, A., Sandoval, F. (eds.) IWANN 2005. LNCS, vol. 3512, pp. 758–770. Springer, Heidelberg (2005). https://doi.org/10.1007/11494669_93
10. Witten, D.M., Tibshirani, R.: A framework for feature selection in clustering. J. Am. Stat. Assoc. **105**(490), 713–726 (2010)
11. Witten, D.M., Tibshirani, R.: Sparcl: perform sparse hierarchical clustering and sparse K-means clustering (2018). https://cran.r-project.org/web/packages/sparcl/

Is t-SNE Becoming the New Self-organizing Map? Similarities and Differences

John A. Lee(✉)

FNRS Research Director Head of Molecular Imaging, Radiotherapy, and Oncology, Université catholique de Louvain, Ottignies-Louvain-la-Neuve, Belgium
john.lee@uclouvain.be

Abstract. Self-organizing maps (SOMs) have many advantages as a tool for exploratory data analysis. Combining vector quantization and topological relationships that are defined in a low-dimensional space, they can run on big data sets and are mostly immune to the curse of dimensionality in the data space.

SOMs are used mainly for dimensionality reduction and marginally for clustering; however, SOMs also suffer from some shortcomings.

Vector quantization makes them unable to embed all data points, only prototypes or centroid are mapped.

Being defined as a regular grid in the low-dimensional space, dimensionality reduction and clustering with SOMs are indirect, as compared to methods of direct embedding like multi-dimensional scaling.

Since 2008, t-SNE (t-distributed stochastic neighbor embedding) has raised growing interest, first in the machine learning community and now outside of it, with many applications in cell biology, for instance.

Primarily used as a 2D embedding and visualization technique, t-SNE is more and more used as a clustering technique, which is capable of identifying meaningful clusters that classical clustering tools struggle to see.

Quite counter-intuitively, t-SNE often better separates clusters in low-dimensional embeddings than clustering tools would do in the high-dimensional data space.

To understand this paradox, several mechanisms of t-SNE can be framed as a distance transformation with the possibility (i) to denoise distances in high-dimensional spaces and (ii) to impose a strong inductive bias on them, which magnifies inter-cluster gaps.

Despite these strengths, t-SNE is not free of drawbacks, which we quickly review to sketch perspectives of future developments.

T. Villmann et al. (Eds.): WSOM+2024, LNNS 1087, p. 42, 2024.
https://doi.org/10.1007/978-3-031-67159-3_5

Pursuing the Perfect Projection: A Projection Pursuit Framework for Deep Learning

Jan-Ole Perschewski(✉), Johann Schmidt, and Sebastian Stober

AILab, Otto von Guericke University, Magdeburg, Germany
{jan-ole.perschewski,johann.schmidt,stober}@ovgu.de

Abstract. The curse of dimensionality refers to phenomena occurring with increasing dimensionality such as marginal differences in distances. Projection pursuit solves this issue by projecting high-dimensional data into a low-dimensional space where meaningful distances allow unbiased function estimation. However, projection pursuit only considers projections onto lines and the unbiased function depends on the sample size. We introduce deep projection pursuit (DPP) to remedy these limitations by using an ensemble of projections on parameterized surfaces combined with neural networks to solve the learning task. Furthermore, we demonstrate the capabilities of the DPP framework by training principal component curves and solving supervised tasks with interpretable models. Finally, we show the ability to maintain group properties in the projection space. Due to these applications, deep projection pursuit is a flexible design paradigm with various use cases.

Keywords: Projection Pursuit · Deep Learning · Interpretability

1 Introduction

The curse of dimensionality are phenomena that occur if data dimensionality becomes large. One phenomenon is that distance functions depending on all features become practically meaningless [3]. For that reason, statistical models that evaluate a distance function such as Kernel Density Estimation or K-Nearest-Neighbour no longer provide meaningful results. Therefore, techniques such as Projection Pursuit (PP) [9] were introduced. Here, the main idea is to project data points into a lower-dimensional space using the scalar product. In this space, distance again becomes meaningful allowing the application of reliant methods. This means we mitigate the impact of the curse of dimensionality while attempting to preserve the data's inherent structure. Additionally, the resulting function can be easily visualized since it only depends on a single variable. In the initial approach by [9], the goal is to find a projection where the projected data is *interesting*. Specifically, the data distribution should be non-Gaussian with locally high density. The general idea of PP was extended for different

T. Villmann et al. (Eds.): WSOM+2024, LNNS 1087, pp. 43–52, 2024.
https://doi.org/10.1007/978-3-031-67159-3_6

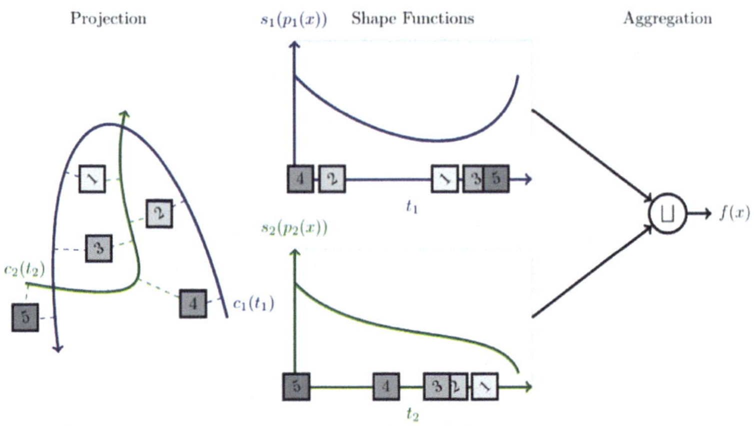

Fig. 1. General Overview of the DPP framework. First, the data (number in squares) is projected onto the underlying curves. Next, the corresponding shape functions are applied leading to a task-specific value. Afterwards, the resulting values are aggregated ($\bigsqcup$) into the overall network output.

tasks such as regression [7], density estimation [8], and classification [19]. These extensions usually include multiple projections, followed by a function from the projection to the target value. Next, the resulting function values are summed or multiplied depending on the task to get the overall model. With this, the general process can be divided into three steps.

First, project data into an *interesting* subspace. Second, determine a function for the given task with respect to the projection and, third, aggregate multiple projections and functions into the overall output. One limitation of these approaches is that they only consider scalar products projecting onto lines or planes. Furthermore, the functions are usually unbiased and therefore too slow to evaluate for data sets with many samples. Hence, we propose to counteract these limitations by combining PP with deep learning. Here, neural networks can approximate arbitrary Riemann-integrable functions and are fast to evaluate on modern hardware. Simultaneously, we keep the idea of low-dimensional projections allowing us to investigate what was learned.

2 Projection Pursuit Framework

The general idea of the Deep Projection Pursuit (DPP) framework is to project data into a space that is helpful to solve a specific task. Ideally, this space has additional beneficial properties, such as the ability to visualize the model depending on a low-dimensional curve or task-dependent biases. Considering the example in Fig. 1, we start with projection onto two curves parameterized by t_i. Afterwards, we apply a generic function on the projected data which solves the given training task to predict the rotation angle and background color. We aggregate the results of the shape function by concatenating the values. Formally, we describe our framework by the following equation:

$$f(x) = \bigsqcup_k s_k(p_k(x_k)) \tag{1}$$

Here, $x_k \in \mathbb{R}^n$ is the input data into the projection. In most cases, we use $x_k = x$ where x is the input data. However, with this formulation, we leave it open to adapt the data for each projection. For instance, we could divide an image into patches and process these patches in parallel. Then again, we could subtract the previous projections building a model that explains away the variances in the data set leading to a sequential dependence on the previous projection. Next, $p_k : \mathbb{R}^n \to \mathbb{R}^n$ is the projection function, and $s_k : \mathbb{R}^n \to \mathbb{R}^l$ is the shape function. Furthermore, $\bigsqcup : \mathbb{R}^l \times \cdots \times \mathbb{R}^l \to \mathbb{R}^m$ aggregates the output of all shape functions with an output-dependent dimensionality l into the final output of the DPP network. In the following, we discuss how to model the defined components.

Projections. The projection function p_k is by definition an idempotent mapping function ($p_k(p_k(x)) = p_k(x)$). Useful projection functions should project the data into a space that has beneficial properties. A typical option for p_k is to project onto a low-dimensional parameterized curve $c : \mathbb{R}^q \to \mathbb{R}^n$ with $q \in \{1, 2\}$. Note, that the dimensionality of the input space q induces the dimensionality reduction. Unfortunately, finding the closest point to x_k on the curve involves solving $t^* = \text{argmin}_{t \in \mathbb{R}^q} ||c(t) - x_k||_2^2$ which does not need to have a closed-form solution. For example, finding the closest point on $c(t) = t^3 \cdot a$ with $a \in \mathbb{R}^n$, already involves a polynomial of degree 5 which has no closed-form solution in terms of radicals. This leads to the necessity to be careful in the choice of the curve and possible approaches to solve the optimization problem. If we can determine the resulting projection $p_k(x_k) = c(t^*)$, we can use the domain of c to visualize where the data is projected and the appearance of the shape function.

Shape Functions. The task of the shape function s_k is to map from the projection space into a task-relevant representation. In the original PP technique [9], the shape function is an unbiased estimator. In contrast, we propose to replace this estimator with an arbitrary neural network. However, it can be helpful to consider the properties of the expected output. For example, we can select a smooth activation function for smooth outputs.

Aggregation. The aggregation function should aggregate the output of all shape functions into an overarching representation. Common choices are summing all entries, taking the product, or concatenation. The choice depends on the given task. For example, the sum is a suitable choice due to the parallels with generalized additive models [13] for regression. However, for density estimation using a product can be beneficial[8].

3 Related Work

The proposed DPP framework can be found or includes many models, which we contrast in the following. **Generalized Additive Models (GAMs)** [13] determine their output by evaluating $f(x) = \beta + \sum_k s_k(x_k)$. Here, $\beta \in \mathbb{R}^l$ is a constant offset and each $s_k : \mathbb{R} \rightarrow \mathbb{R}^l$ is a function mapping the value of a single variable into the output space. This means a GAM is DPP model with $p_k(x) = x_k$ which is a projection onto the unit vectors. GAMs are a widely popular approach in machine learning [14,21,23]. Moreover, some approaches use neural networks as s_k in an GAM such as the work by [1] presenting Neural Additive Models (NAMs). [27] extend the approach by adding the sum $\sum_{(i,j)} s(x_i, x_j)$ which allows modeling pairwise interaction. This extension is also included in the DPP framework using $p_{i,j}(x) = \begin{pmatrix} x_i & x_j \end{pmatrix}$ Similarly, [22] improve the efficiency of NAMs using a weighted sum of shape functions. **Index Models** [25] evaluate $f(x) = g(\theta_1^T x, \theta_2^T x, ..., \theta_k^T x)$. Here, we can distinguish between Single-Index and Multi-Index models where $k = 1$ and $k > 1$, respectively. In this case, $g : \mathbb{R}^k \rightarrow \mathbb{R}^l$ combines the shape function and the aggregation operator. In this sense, this formulation is more general, however, the projection is fixed to the result of a scalar product. Which excludes potentially interesting projections from this framework. Existing works employ shallow neural networks [2,4] within this framework, but refrain from increasing the depth due to the theoretical focus. **Autoencoders** [24] are used to reduce the dimensionality of data. They consist of an encoder $e : \mathbb{R}^n \rightarrow \mathbb{R}^q$ that reduces a feature vector, and a decoder $d : \mathbb{R}^q \rightarrow \mathbb{R}^n$ that reconstructs the original input. Autoencoders create low-dimensional representations if $q \in \{1, 2\}$. Analogously, we have the interpretable projection if the underlying curve in DPP is low-dimensional. The main difference between both low-dimensional representations is that the reduction in the autoencoder is trained whereas the DPP approach searches for the low-dimensional representation.

4 Experiments

In the following, we show the capabilities of models following the DPP framework. We start with the training of principal component curves, followed by the training of a regression model. Next, we investigate the latent space for group properties.

4.1 Principal Component Curves

The DPP framework can model principal curves [12]. Principal curves are an extension of principal components that include nonlinearity within the data. The original approach iteratively creates a function fitting the data. Similar approaches are elastic net [10] or self-organizing maps [18] which adapt control points within a low-dimensional geometry. In contrast, we use a single neural network $N(t;\theta) : \mathbb{R} \rightarrow \mathbb{R}^n$ to represent the low-dimensional curve which allows us to ignore the aggregation. However, the projection onto this curve is not trivially due to the nonconvexity and the unavailable closed-form solution. Hence, we project x onto N by applying gradient descent [5] to solve $\text{argmin}_{t\in\mathbb{R}} ||N(t;\theta) - x||^2$. Applying gradient descent trivially results in a large computational graph due to the repeated evaluation of the derivative. To mitigate that issue, we search locally for the optimal step size using an initial line search for the end of a local minimum followed by a binary search for refinement. As a shape function, we select the identity since we want the projection to fit the data manifold. The loss function for our Neural Network is the Mean-Square Error over a batch with size M weighted by the closeness to the curve $\frac{1}{M}\sum_{i=1}^{M}\max(e^{-||p(x_i)-x_i||_2^2}, \varepsilon)||p(x_i) - x_i||_2^2$. We include the closeness of the curve into the loss to keep the network in high-density regions. If the curve is in a high-density region, the curve is kept there due to the higher weight. If the curve is in a low-density region it is pulled towards a close data point due to the lower bound of $\varepsilon = 0.05$ on the MSE. The value of ε needs to be selected depending on the task. Empirically, we observed that otherwise gaps between dense regions are preferred. In this experiment, we fit a simple data set analogously to [12]. This means, we sample 100 points from a unit circle regularly and add Gaussian noise with a variance of 0.1. The DPP model starts with restricting t between $(-1, 1)$ by applying the *tanh* function followed by fully-connected layers with 128, 128, and 2 neurons. After each but the last layer, we apply the GELU [15] activation, which results in a smooth curve. We train the model with all samples for 900 steps using the Adam optimizer with a learning rate of 0.001, $\beta_1 = 0.9$, and $\beta_2 = 0.999$ which we will refer to as default parameters [17]. During the training depicted in Fig. 2, we can see that the neural network curve fits the generated data. This shows that projections onto neural network curves are feasible and fit the underlying structure.

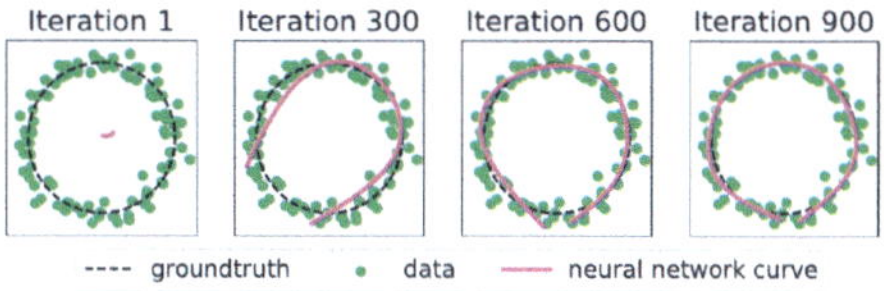

Fig. 2. During training, the curve parameterized by a neural network fits the underlying structure successfully.

4.2 Reduced Neural-Additive Models

In this experiment, we demonstrate that DPP can lead to interpretable models that are more compact than NAMs. Hence, we repeat an experiment from [1] using the California housing data [20]. Here, the task is to predict the median house value within a block in \$100.000 using eight features for each block: median income, median house age, average number of rooms and bedrooms, population size, number of household members, longitude and latitude of the block.

For the experiment, we project the data onto multiple lines parameterized by $t_k \in \mathbb{R}$ as $c_k(t_k) = o_k + t_k \cdot b_k$ with $o_k, b_k \in \mathbb{R}^8$ the position vector and direction vector, respectively. In this case, the optimization $t_k^* = \text{argmin}_t ||o_k + t_k \cdot b_k - x||_2^2 = (x^T b_k - o_k^T b_k)/b_k^T b_k$ has a closed-form solution. Furthermore, we can simplify the projection by reparameterizing $b_k = \tilde{b}_k/||\tilde{b}_k||_2$ leading to $t_k^* = (x^T b_k - o_k^T b_k)$. Each of the k projections is applied to the standardized input vector. Furthermore, each shape function consists of a three-layer neural network with 64, 64, and one neuron for the prediction analogously to [1]. We also apply the ReLU activation after the layers with 64 neurons. Finally, we use the sum of all outputs as the aggregation function if the number of projections exceeds one.

For training, we apply the same protocol as [1]. Specifically, we performed 5-fold cross-validation with 20 combinations of 80% training data and 20% testing data. We also randomly split the 80% training data to use 70% of the data for training and 10% for early stopping. For optimization, we minimize RMSE loss using full batch training for 20000 steps employing the Adam optimizer with default parameters checking the validation performance every 500 steps.

Table 1. Comparison of performance on California housing data analogously to [1](*) for (left) different number of projections and (right) comparable approaches.

Projections	$RMSE(\pm std)$
1	0.643 ± 0.076
2	0.608 ± 0.019
3	0.570 ± 0.015
4	0.566 ± 0.025
5	0.544 ± 0.024
6	0.541 ± 0.025
7	0.533 ± 0.018
8	0.529 ± 0.022

Approach	RMSE($\pm std$)
DNNs*	0.492 ± 0.009
DPP(8 Projections)	0.529 ± 0.022
XGBoost*	0.532 ± 0.014
EBMs*	0.557 ± 0.009
NAMs*	0.562 ± 0.007
DPP(4 Projections)	0.566 ± 0.025
CART*	0.720 ± 0.006
Log./Linear Reg*	0.728 ± 0.015

First, we investigate the performance regarding the number of projections in the described model. We expect to see a lower error with an increasing number of projections due to the increase in the number of parameters and larger bottleneck. In our experiment, we observe the expected behavior considering the errors summarized in Table 1(left) If we consider the performance in comparison to the results reported by [1] in Table 1(right), we can see that the performance of our DPP model with four projections is close to the performance of the NAM. This is especially surprising since the DPP model applying four projections uses 31932 fewer parameters. This shows that the DPP model is able to achieve a similar performance whilst keeping the number of parameters small.

Next, we introduce a visualization of a trained model in Fig. 3. Here, we see the results of the best model (0.52 RMSE) with four projections from the 20 runs performed for Table 1. For the visualization, we combine a table with a plot showing the value $s_k(c_k(t_k))$ with respect to t_k. Here, the range of t is chosen to include 98% of the data excluding the highest and lowest percent to mitigate the influence of the outliers on the visualization. The background of the plot indicates the density of the data in the specific region. For that, we created a histogram of the curve parameters t_k^* with respect to the data set and divided the counts in each bin by the maximum count.

The corresponding table shows the Zero-Values and One-Step values. The Zero-Values correspond to the learned position vector and show the features

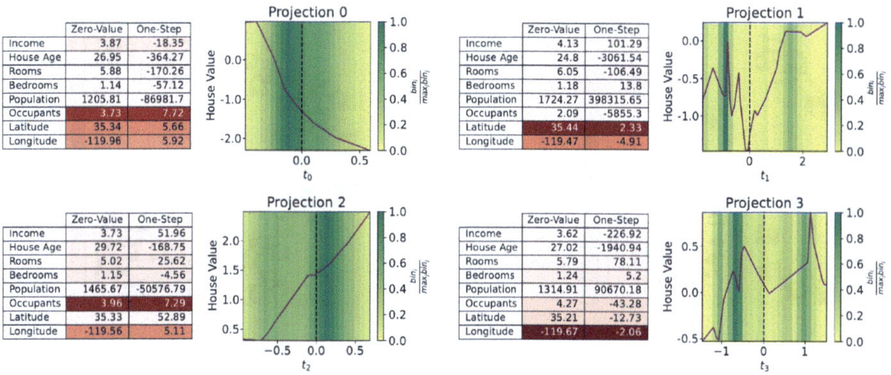

Fig. 3. Visualization of the best DPP model using four projections. For each projection, we show a table on the left defining the values for the Zero-Value and One-Step and a plot showing the house pricing depending on the value t_k from the projection. The Zero-Value is the location vector of the projection corresponding to $t_k = 0$. The One-Step values show how much we need to add to this quantity to increase t_k by one.

such that $t_k = 0$. The One-Step values show the influence of a specific attribute on the value of t_k. Specifically, it answers how much we need to add to a single feature to increase t_k by 1. Hence, if the One-Step value is large the variable is less important for this curve. Additionally, we indicated the importance of a feature by the darkness of the underlying color so that the light rows can be ignored while interpreting the plot.

We investigate the model with four projections in Fig. 3. At first glance, we can see that multiple features can be ignored due to the magnitude of the One-Step such as income, house age, and population. In projections 0 and 2, we both have a strong dependency on the number of occupants, which adds up to a slight decrease in value with an increase in occupants. Projection 1 and projection 3, can be interpreted to identify regions with high house values which correspond to large cities near the coast such as San Diego (32″N, −117″O).

4.3 DPP Preserves Group Structure

The subtask of projecting onto a curve often aligns with the objective of manifold learning algorithms[16]. Here, the objective is to preserve structure when mapping $\mathbb{R}^n \rightarrow \mathbb{R}^m$. Following the Manifold Hypothesis [6] we assume $n \gg m$. For visualization purposes, a common choice is $m = 2$. However, the data manifold embedded in $\mathbb{R}^n$ is unknown. Under this light, we only consider the known Lie group [11] of center-rotations, $\mathtt{SO}(2)$, which is continuous and infinite. Each element in this group $g \in \mathtt{SO}(2)$ is represented by an orthogonal matrix of the form $\left[\begin{smallmatrix}\cos(\theta) & -\sin(\theta)\\ \sin(\theta) & \cos(\theta)\end{smallmatrix}\right]$ with some rotation angle $\theta \in \mathbb{R}$.

Instead of experimenting with these continuous structures, we consider finite subgroups instead, i.e., $\langle g \rangle = \{g^k \mid k \in \mathbb{Z}\}$ generated from G. That means we

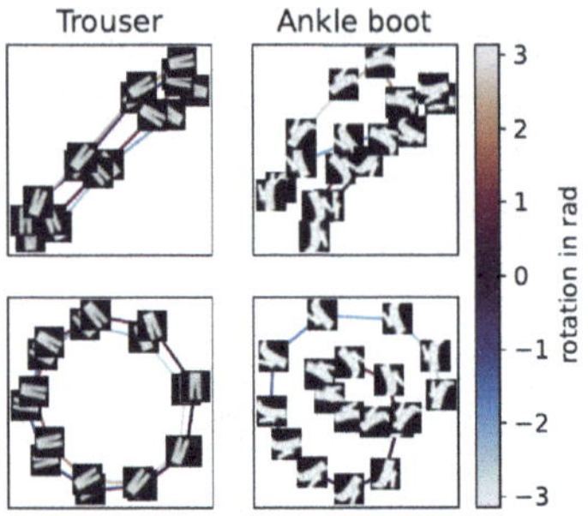

(a) DPP

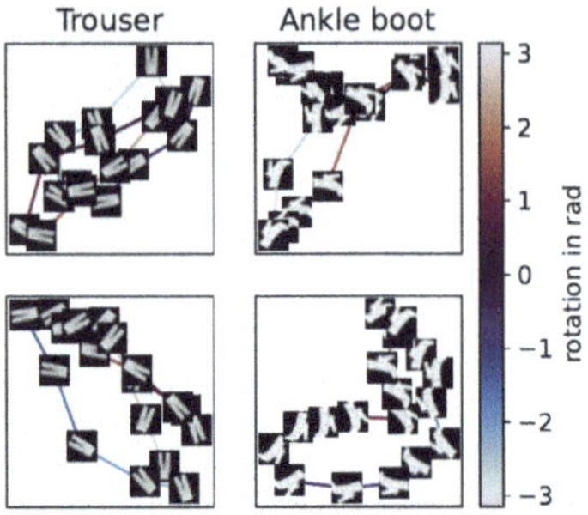

(b) Autoencoder

Fig. 4. Orbit of images in the latent spaces for a full center rotation in 20° steps. In (a), we see the trajectory with respect to the curve parameters for the two distinct curves showing cyclic structures. In (b), we see the corresponding latent spaces in a similar autoencoder with less regular trajectories.

consider finite rotation angles, e.g., $\pi/2, 2\pi/2, \ldots, k\pi/2$. For any image $\mathbf{x}$ this forms an orbit, $\text{orb}(\langle g\rangle, \mathbf{x}) = \{g\mathbf{x} \mid g \in \langle g\rangle\}$. This finite-sized object allows measuring how well DPP preserves structure. The orbit for every $\mathbf{x}$ is cyclic, i.e., all nearest neighbors are equally distant. This can be thought of as the approximation of a circle by a regular polygon of size $|\langle g\rangle|$. A mapping function, that preserves these orbits is isomorphic to G[11], which is a desirable property for robust projections. In summary, the projection of elements from the ambient data manifold in $\mathbb{R}^n$ onto the latent information manifold in $\mathbb{R}^m$ is structure-preserving if all group orbits maintain their geometric and topological properties. That means they keep equally distant nearest neighbors in a cycle.

In this experiment, we investigate a DPP model with two projections. Each projection consists of a plane $c_k(t_1, t_2) = o + t_1 \cdot b_1 + t_2 \cdot b_2$. To solve the optimization problem $||c(t_1, t_2) - x||_2^2$, we need to ensure, that b_1 and b_2 are orthonormal allowing a closed-form solution. For that, we ensure orthonormality by reparameterizing $b_1 = \tilde{b}_1/||\tilde{b}_1||$ and $b_2 = \tilde{b}_2/||\tilde{b}_2||$ where $\tilde{b}_2 = \hat{b}_2 - (\tilde{b}_1^T \hat{b}_2)/(\tilde{b}_1^T \tilde{b}_1)\tilde{b}_1$ which is a single step of the Gram-Schmidt process. After the projection, we apply three fully-connected layers with 500, 500, and 784 neurons with ReLU between all layers as the shape function, followed by summation as aggregation. For comparison, we use a similar autoencoder with a linear layer reducing the input to a 2 by 2 representation. We increase the dimensionality using a linear layer with 784 units for each pair. Next, we use the same architecture as the shape function in the DPP model and sum the shapes. Hence, the main difference is that the latent space of the DPP model is restricted to two planes, whereas the autoencoder is free to use the 2 by 2 latent space as required. We train both models on the FashionMNIST [26] data set with random rotation of up to 35° for three epochs with a batch size of 512. Using the Adam optimizer with default parameters we minimize the Mean Squared Error (MSE) between reconstruction and input.

After training, we compare the obtained projections. Ideally, the samples are along the orbit equally distant. Our qualitative results are depicted in Fig. 4. The DPP model can preserve the group structure by being noticeably closer to circle structures than the autoencoder's embedding. Although these global orbit structures emerge, we can observe local structures that seem to cluster images with small pixel-wise distances. See, for instance, the lower trouser plot in Fig. 4(a), where we can see that the angle of the trousers aligns, indicating that the orientation of the trousers is lost due to the usage of the MSE.

We support this initial finding by quantitatively analyzing the average obtained structures. To remove scaling dependencies, we start by standardizing the data. We measure the pairwise Euclidean distances based on the orbit structure, $\|g^k f(\mathbf{x}) - g^{k+1} f(\mathbf{x})\|_2$. We estimate the expected variance along an orbit for all test orbits within a two-dimensional subspace. Next, we take the minimum of the two measures since the appearance in one of them is sufficient. The resulting scalar measures how well the projection manifold preserves the structure of $\mathtt{SO}(2)$, which would have no variance. For comparison, we average the variances of 10 models of each type. Our DPP model achieves a score of $0.002 \pm 6e-4$, half of the autoencoder ($0.005 \pm 8e-4$). This shows that a DPP model preserves group structure better than a comparable autoencoder.

5 Discussion

In this paper, we proposed to generalize projection pursuit into the deep projection pursuit framework by combining generic projections, shape functions, and aggregations. First, we demonstrate the capability to find principal component curves by finding projections onto nonlinear surfaces using gradient descent. Furthermore, we showed that the DPP framework can be used to train interpretable regression models that perform feature selection and achieve a competitive performance to similar models. Finally, we discovered that a projection pursuit autoencoder leads to discoverable group properties given a data set including small rotations. Overall, we see that the DPP framework is a flexible framework applicable to various tasks. However, the current experiments are still small-scale, and scaling them to more complex data such as high-resolution or color images are important research tasks yet to be addressed.

References

1. Agarwal, R., et al.: Neural additive models: interpretable machine learning with neural nets. In: NeurIPS, vol. 34, pp. 4699–4711. Curran Associates, Inc. (2021)
2. Bach, F.: Breaking the curse of dimensionality with convex neural networks. J. Mach. Learn. Res. **18**(1), 629–681 (2017)
3. Beyer, K., Goldstein, J., Ramakrishnan, R., Shaft, U.: When is "nearest neighbor" meaningful? In: Beeri, C., Buneman, P. (eds.) ICDT 1999. LNCS, vol. 1540, pp. 217–235. Springer, Heidelberg (1999). https://doi.org/10.1007/3-540-49257-7_15
4. Bietti, A., Bruna, J., Sanford, C., Song, M.J.: Learning single-index models with shallow neural networks. In: NeurIPS, vol. 35, pp. 9768–9783. Curran Associates, Inc. (2022)

5. Domke, J.: Generic methods for optimization-based modeling. In: 15.AISTATS. Proceedings of Machine Learning Research, La Palma, Canary Islands, vol. 22, pp. 318–326. PMLR (2012)
6. Fefferman, C., Mitter, S., Narayanan, H.: Testing the manifold hypothesis. J. Am. Math. Soc. **29** (2016). https://doi.org/10.1090/jams/852
7. Friedman, J.H., Stuetzle, W.: Projection pursuit regression. J. Am. Stat. Assoc. **76**(376), 817–823 (1981)
8. Friedman, J.H., Stuetzle, W., Schroeder, A.: Projection pursuit density estimation. J. Am. Stat. Assoc. **79**(387), 599–608 (1984)
9. Friedman, J., Tukey, J.: A projection pursuit algorithm for exploratory data analysis. IEEE Trans. Comput. **C-23**(9), 881–890 (1974). https://doi.org/10.1109/T-C.1974.224051
10. Gorban, A., Rossiev, A.: Neural network iterative method of principal curves for data with gaps. J. Comput. Syst. Sci. Int. **38**(5), 825–830 (1999)
11. Hall, B.C.: Lie Groups, Lie Algebras, and Representations. GTM, vol. 222. Springer, Cham (2015). https://doi.org/10.1007/978-3-319-13467-3
12. Hastie, T., Stuetzle, W.: Principal curves. J. Am. Stat. Assoc. **84**(406), 502–516 (1989). https://doi.org/10.1080/01621459.1989.10478797
13. Hastie, T., Tibshirani, R.: Generalized additive models. Stat. Sci. **1**(3), 297–310 (1986). https://doi.org/10.1214/ss/1177013604
14. Hastie, T., Tibshirani, R.: Generalized additive models for medical research. Stat. Methods Med. Res. **4**(3), 187–196 (1995). https://doi.org/10.1177/096228029500400302
15. Hendrycks, D., Gimpel, K.: Bridging nonlinearities and stochastic regularizers with gaussian error linear units. CoRR abs/1606.08415 (2016)
16. Izenman, A.: Introduction to manifold learning. Wiley Interdisc. Rev.: Computat. Stat. **4** (2012). https://doi.org/10.1002/wics.1222
17. Kingma, D.P., Ba, J.: Adam: a method for stochastic optimization. In: Bengio, Y., LeCun, Y. (eds.) 3rd International Conference on Learning Representations, ICLR 2015, San Diego, CA, USA, 7–9 May 2015, Conference Track Proceedings (2015)
18. Kohonen, T.: The self-organizing map. Proc. IEEE **78**(9), 1464–1480 (1990). https://doi.org/10.1109/5.58325
19. Lee, E.K., Cook, D., Klinke, S., Lumley, T.: Projection pursuit for exploratory supervised classification. J. Comput. Graph. Stat. **14**(4), 831–846 (2005)
20. Pace, R.K., Barry, R.: Sparse spatial autoregressions. Stat. Probab. Lett. **33**(3), 291–297 (1997)
21. Pedersen, E.J., Miller, D.L., Simpson, G.L., Ross, N.: Hierarchical generalized additive models in ecology: an introduction with MGCV. PeerJ **7**, e6876 (2019)
22. Radenovic, F., Dubey, A., Mahajan, D.: Neural basis models for interpretability. In: NeurIPS, vol. 35, pp. 8414–8426. Curran Associates, Inc. (2022)
23. Ravindra, K., Rattan, P., Mor, S., Aggarwal, A.N.: Generalized additive models: building evidence of air pollution, climate change and human health. Environ. Int. **132**, 104987 (2019). https://doi.org/10.1016/j.envint.2019.104987
24. Rumelhart, D.E., Hinton, G.E., Williams, R.J.: Learning Internal Representations by Error Propagation, pp. 318–362. MIT Press, Cambridge (1986)
25. Stoker, T.M.: Consistent estimation of scaled coefficients. Econometrica **54**(6), 1461–1481 (1986)
26. Xiao, H., Rasul, K., Vollgraf, R.: Fashion-MNIST: a novel image dataset for benchmarking machine learning algorithms (2017)
27. Yang, Z., Zhang, A., Sudjianto, A.: GAMI-Net: an explainable neural network based on generalized additive models with structured interactions. Pattern Recogn. **120**, 108192 (2021). https://doi.org/10.1016/j.patcog.2021.108192

Generalizing Self-organizing Maps: Large-Scale Training of GMMs and Applications in Data Science

Alexander Gepperth(✉)

Fulda University of Applied Sciences, Leipzigerstr. 123, 36037 Fulda, Germany
alexander.gepperth@cs.hs-fulda.de
https://www.gepperth.net/alexander

Abstract. This contribution shows that Gaussian Mixture Models can be considered generalizations of self-organizing maps. More precisely, we demonstrate that the training of self-organizing maps is an approximation to the training of Gaussian Mixture Models by gradient descent. As a consequence, the scores of a trained SOM can be treated as log-likelihoods of a GMM with tied, spherical covariance and used, e.g., for outlier detection, whereas sampling from trained SOMs is not well-defined. Furthermore, we outline how SGD-trained GMMs can be generalized to diagonal and more expressive covariance matrices and how this benefits typical data science applications such as outlier detection, sampling and generative classification. Source codes are available on the author's web site or upon request.

Keywords: Self-Organizing Maps · Gaussian Mixture Models · Stochastic Gradient Descent

1 Introduction

The self-organizing map (SOM) model [5] has widely been used in data visualization and data exploration, and has been very influential in general. SOM training is however complicated by the fact that there is no loss function that is minimized by SOM training, and that the decay of the neighbourhood radius over time is tricky to tune. In addition, is it theoretically unclear how SOM scores, i.e., input-prototype distances, can be interpreted although they have been used for outlier detection on an ad-hoc basis.

This article proposes the Gaussian Mixture Model (GMM) as a "generalized self-organizing map", i.e., a probabilistic model that reduces to SOM in a suitable approximation. By virtue of this approximation, one can harness the beneficial properties of SOMs while at the same time providing SOM with a clear probabilistic interpretation, leading to applications in outlier detection and sampling with SOM-like models.

T. Villmann et al. (Eds.): WSOM+2024, LNNS 1087, pp. 53–62, 2024.
https://doi.org/10.1007/978-3-031-67159-3_7

1.1 Related Work

SOMs and GMMs have multiple common points, the most notable being that they are based on *distances* rather than scalar products as DNNs are, which is why some authors include them in the group of *prototype-based models*. There are a few works that try to link the GMM and SOM models in the past. Notably, in [3], the authors propose a modified SOM model that is based on the minimization of a differentiable loss. This was new at the time, since it had been shown that the SOM learning rule cannot be derived from the minimization of *any* single differentiable loss function, see [4]. On the other hand, is was demonstrated in [9] that SOM-like topological visualization behavior can be imposed on GMMs if the common uniform prior over latent variables is replaced appropriately. However, gradient-based training of GMMs is not discussed in this article, and neither is a formal link between SOMs and GMMs. The possibilities of advanced GMMs such as the MFA model for image modeling were explored in [8], using SGD for using k-means initialization.

1.2 Contributions

This article makes the following contributions:

- mathematical proof that the SOM learning rule is an approximation to the training of GMMs by stochastic gradient descent
- provide a probabilistic analysis of SOM's ability to sample and perform outlier detection
- demonstrate the capabilities of SGD-trained GMMs as "generalized SOMs" w.r.t. sampling, outlier detection and generative classification

2 Theoretical Contributions

2.1 Relation Between GMM and SOM Training

Gaussian Mixture Models. The GMM model approximates an unknown distribution, given empirically by data samples $X = \{\boldsymbol{x}_n\}$, by expressing that distribution by a weighted superposition of K multivariate Gaussian densities $\mathcal{N}_k(\boldsymbol{x}) \equiv \mathcal{N}(\boldsymbol{x}; \boldsymbol{\mu}_k, \boldsymbol{\Sigma}_k)$: $p(X) = \prod_n \left(\sum_k \pi_k \mathcal{N}_k(\boldsymbol{x}_n)\right)$. When passing to log-probabilities, this transforms to $\mathcal{L}_0 = \log p(X) = \sum_n \log \sum_k \pi_k \mathcal{N}_k(\boldsymbol{x}_n)$, which is often termed the GMM log-likelihood. GMMs are trained by adapting the *weights* π_k, the *centroids* $\boldsymbol{\mu}_k$ and the covariance matrices $\boldsymbol{\Sigma}_k$ such that the negative log-likelihood is maximized.

GMM Optimization by SGD. Traditionally, GMMs are trained by the Expectation-Maximization (EM) algorithm which does not require learning rates and has good convergence properties. However, since EM is inherently a batch-type algorithm, i.e., it needs to process all samples at once to retain its convergence properties, it is unfeasible for large-scale training. Furthermore, convergence in practice is only ensured if centroids are initialized by, e.g., k-means.

This is why recent publications [2] propose to optimize GMMs by SGD from random initial conditions. In order for this, two modifications need to be made: first of all, the log sum inside the GMM log-likelihood is approximated by its largest term:

$$\begin{aligned}\mathcal{L}_1(X) = \mathcal{L}_0(X) &= \sum_n \log \sum_k \pi_k \mathcal{N}_k(\boldsymbol{x}_n) \approx \sum_n \log \max_k \pi_k \mathcal{N}_k(\boldsymbol{x}_n) \\ &= \sum_n \max_k \log\left(\pi_k \mathcal{N}_k(\boldsymbol{x}_n)\right)\end{aligned} \tag{1}$$

This is very numerically advantageous as the log removes the exponential inside the Gaussian density $\mathcal{N}$, but it also makes the gradients much more tractable. When optimizing this loss by SGD from random initial conditions, we often encounter *degenerate solutions* where one component describes all the data, and the remaining ones remain uninitialized. To remedy this, [2] proposes to add an annealing mechanism to the loss of Eq. (1) which averages components over a Gaussian neighbourhood, thus obtaining the final GMM loss

$$\begin{aligned}\mathcal{L}_2(X) = \mathcal{L}_1(X) &= \sum_n \max_k \log\left(\pi_k \mathcal{N}_k(\boldsymbol{x}_n)\right) \\ &\approx \sum_n \max_k \left[\sum_l g_{kl}(\sigma) \log\left(\pi_l \mathcal{N}_l(\boldsymbol{x}_n)\right)\right] \equiv \mathcal{L}\end{aligned} \tag{2}$$

The kernel matrix g_{ij} follows a Gaussian profile with a time-dependent width $\sigma(t) \equiv \sigma$: $g_{kl} = \exp(-0.5 d(k,l)^2 \sigma^{-2})$, with $d(i,j)$ representing the Euclidean distance between components i and j when placed on a two-dimensional grid. It is easy to see that $g_{kl} \rightarrow \delta_{kl}$ in the limit $\sigma \rightarrow 0$, in which case the annealed loss $\mathcal{L}_2$ reduces to $\mathcal{L}_1$. Optimizing a GMM by SGD is therefore performed by maximizing $\mathcal{L}_2(X)$ w.r.t $\boldsymbol{\mu}_k, \boldsymbol{\Sigma}_k, \pi_k$ while letting $\sigma(t)$ decay from some initial value σ_0 to 0. This ensures that local optima are avoided by the optimization, as demonstrated and proven in [2]. An additional reparameterization ensures that weights are normalized: $\sum_k \pi_k = 1$.

From GMM to SOM. In the GMM model, the covariance matrices $\boldsymbol{\Sigma}_k$ can have any form. Most commonly, they are chosen as spherical ($\boldsymbol{\Sigma}_k = \sigma_k \boldsymbol{I}$) or diagonal matrices ($\boldsymbol{\Sigma}_k = \text{diag}[\sigma_{k1}, \sigma_{k2}, dots]$)) for simplicity. If we assume that all covariance matrices have a spherical, tied form of unit variance as $\boldsymbol{\Sigma}_k = I$, the covariance matrices effectively drop out of the GMM loss $\mathcal{L}_2$. The Gaussian component densities $\mathcal{N}_k$ thus simplify as

$$\begin{aligned}\mathcal{N}_k &\equiv \mathcal{N}(\boldsymbol{x}; \boldsymbol{\mu}_k, \boldsymbol{\Sigma}_k) = (2\pi \det \boldsymbol{\Sigma}_k)^{-0.5} \exp(-0.5\left((\boldsymbol{x} - \boldsymbol{\mu}_k)^T \boldsymbol{\Sigma}_k^{-1} (\boldsymbol{x} - \boldsymbol{\mu}_k)\right) \\ &\rightarrow \sqrt{2\pi} \exp(-0.5 ||\boldsymbol{x} - \boldsymbol{\mu}_k||^2).\end{aligned} \tag{3}$$

If we furthermore assume that the weights are all tied to K^{-1}, they drop out of the loss as well, which finally becomes:

$$\begin{aligned}\mathcal{L}_2 &= \sum_n \max_k \left[\sum_l g_{kl}(\sigma)\log\left(\pi_l \mathcal{N}_l(\boldsymbol{x}_n)\right)\right] \\ &\rightarrow \sum_n \max_k \left[\sum_l g_{kl}(\sigma)(-0.5\log(2\pi) - 0.5||\boldsymbol{x}_n - \boldsymbol{\mu}_k||^2)\right] \\ &= -0.5\sum_n \max_k \left[\sum_l g_{kl}(\sigma)(||\boldsymbol{x}_n - \boldsymbol{\mu}_k||^2)\right] + \text{const.} \equiv \mathcal{L}_3 \end{aligned} \tag{4}$$

Discarding the constant and the leading factor which are irrelevant for gradient descent, we arrive at what was termed the "energy function" for the SOM model proposed by [3]. When optimizing $\mathcal{L}_3$ by stochastic gradient descent, we obtain the update rule

$$\boldsymbol{\mu}_l \leftarrow \boldsymbol{\mu}_{k^*} + \epsilon g_{k^*l}(\sigma)(\boldsymbol{\mu}_{k^*} - \boldsymbol{x}), \tag{5}$$

which of course has to be accompanied by an appropriate decrease regime for $\sigma = \sigma(t)$. The only difference to the conventional online SOM learning rule is that the best-matching unit (BMU) k^* is determined here as $k^* = \arg\max_k \left[\sum_l g_{kl}(\sigma)(||\boldsymbol{x}_n - \boldsymbol{\mu}_k||^2)\right]$, instead of $k^* = \arg\max_k ||\boldsymbol{x}_n - \boldsymbol{\mu}_k||$. In the limit of $\sigma \rightarrow 0$, these two formulations become identical. For all practical purposes, the modified BMU selection leads to the same results, as shown in [4].

2.2 Analysis of SOM Capabilities

With the knowledge that SOMs are a particular limiting case of GMMs, we can derive the following statements w.r.t. particular SOM capabilities:

SOMs Can Perform Limited Outlier Detection. SOM scores $s_k \equiv -||vx - \boldsymbol{\mu}_k||^2$ correspond to GMM log-probabilities, for GMMs with tied weights and spherical unit covariance, see above. We can therefore compute the posterior probability that a given sample was produced by a component c as $p(\boldsymbol{x}|c) = \frac{\exp(s_c)}{\sum_l \exp(s_l)}$ and impose a threshold for the BMU probability. Alternatively, we can apply an application-derived threshold directly on the SOM scores to detect outliers. This is of course only possible under the assumption that data clusters do have approximately unit variance.

SOMs Cannot be Used for Sampling. For outlier detection, it is immaterial whether the unit variances fit the actual data well, since we are just want to rank SOM's components w.r.t. their degree of match to the data. For sampling, this is no longer true, since we need the variances to draw samples from a multivariate Gaussian distribution. By assuming unit variances for all components, all clusters in sampled data will therefore have unit radius regardless of their actual extent in data space. Thus, SOMs are not suitable for faithful sampling.

2.3 Advantages of GMMs w.r.t SOMs

GMMs do not share the limitations of SOMs outlined in Sect. 2.2, and if they are trained by SGD, they are by construction capable of outlier detection and sampling. However, they do retain many important SOM properties, while at the same time fixing a serious issue in SOM training:

Automatic Tuning of Neighbourhood Radius. The correct decay regime for $\sigma(t)$ is always hard to tune for SOMs because there is no real criterion of SOM performance (since there is not loss function to be minimized). In contrast, SGD-trained GMMs provide a loss function and thus a simple criterion for reducing $\sigma(t)$: essentially, this should happen whenever the loss is stationary. It can be shown (see [2]) that σ defines a lower bound for the GMM loss $\mathcal{L}_2$, so reducing σ can allow the loss to decrease further. Performing $\sigma \rightarrow \gamma\sigma$ in case of stationary loss during training provides a viable strategy that does not require manual tuning at all. The only values to set are $\sigma_0 \equiv \sigma(t = 0)$ and σ_∞, both of which can be set based on heuristics.

Data Visualization Property. Since the loss $\mathcal{L}_2$ rewards configurations where similar centroids lie close together on the two-dimensional lattice, SGD-trained GMMs that optimize $\mathcal{L}_2$ retain the topological SOM property and can thus be used for data visualization and exploration as a drop-in replacement for SOMs.

Arbitrary Covariance Matrices. One particular feature of GMMs is the freedom to choose a particular form for the covariance matrices. In practice, diagonal CMs are very common especially for high-dimensional data, since GMM training involves matrix inversions, which makes full covariance matrices too expensive in terms of memory and computation time. An interpolation between these two choices for a d-dimensional problem is given by the Mixture of Factor Analyzers (MFA) model which decomposes the CM as $\boldsymbol{\Sigma} = \boldsymbol{D} + \Gamma\Gamma^T$, where $\boldsymbol{D} = \mathrm{diag}(\sigma_1, \ldots, \sigma_d)$ and $\Gamma \in \mathbb{R}^{l \times d}$. The parameter $l \in [1, d]$ determines the complexity of the resulting CM. Just like GMMs, MFA instances can be trained by SGD, see [1].

3 Experiments

In all experiments of this section, we use GMM and SOMs with $K = 64$ components and periodic boundary conditions on a rectangular lattice. The initial neighbourhood radius for SOM and GMM experiments is set to a quarter of the diagonal of the lattice, i.e., $\sigma_0 = \sqrt{128}/4 \approx 3$. The asymptotic neighbourhood radius is set to $\sigma_\infty = 0.01$. For SOM experiments, the decay of the neighbourhood radius is performed as $\sigma = \sigma_0 \exp(-e)$, where $e > 1$ denotes the current epoch. For $E = 1$, the neighbourhood radius is kept at σ_0. For GMM training by SGD, the automatic annealing mechanism is used for controlling the "neighbourhood radius" $\sigma(t)$, with default parameters as given in [2].

We use the following image datasets for our experiments. All of them are relatively high-dimensional and well out of reach of conventional EM-trained GMMs.

MNIST. [6] consists of 60.000 28 × 28 grayscale images of handwritten digits (0–9).

Fashion-MNIST. [10] consists of 60.000 images of clothes in 10 categories and is structured like MNIST.

Fruits-360 [7] contains 100 × 100 images showing different types of fruits, from which we chose the 10 best-represented classes and downsample to 48 × 48 RGB.

3.1 Data Visualization with SOMs, Energy-Based SOMs and GMMs

Fig. 1. Centroids obtained after training a SOM (left), energy-based SOM (middle) and GMM (right), arranged on a 10 × 10 grid, after training on the Fruits-360 dataset. We observe a similar topological organization in all three cases.

This is a preliminary experiment meant to show that the data visualization property of SOMs actually carries over to GMMs, in this case with a diagonal covariance matrix. The $K = 100$ GMM centroids, obtained after 100 training epochs on the Fruits-360 dataset, are visualized in Fig. 1, and we observe that they show the SOM-like topological organization where similar prototypes/centroids develop next to each other during training.

3.2 Conditional and Unconditional Sampling from GMMs

Table 1. Classification accuracies for linear classification based on MFA and GMM responsibilities on various datasets. We always use $K = 64$ components for GMM and MFA and use two values for the MFA latent dimension l. All accuracies are measured on the test sets of the respective datasets.

dataset	Fruits-360	MNIST	FashionMNIST
GMM	98.6	89.3	85.2
MFA, $l = 4$	98.6	89.9	88.8
MFA, $l = 25$	99.0	91.7	90.9

In this section, we demonstrate sampling from GMMs with diagonal covariance matrices, which is performed by assuming an uniform prior over components and drawing components from it. In addition, we show that coupling the responsibilities of a GMM to a simple linear classifier allows for the conditional sampling of given classes. The logic behind this approach is simple: the normalized GMM/MFA responsibilities (or posterior probabilities) $\gamma_k(\boldsymbol{x}) \equiv \frac{p_k(\boldsymbol{x})}{\sum_i p_i(\boldsymbol{x})}$ are transformed as $\boldsymbol{y} = \boldsymbol{S}(W\gamma(\boldsymbol{x}) + \boldsymbol{b})$, where $\boldsymbol{S}(\boldsymbol{y})$ denotes the softmax function. After being trained by cross-entropy, this linear classifier can be approximately inverted as $\hat{\gamma} = W^T(\boldsymbol{c} - \boldsymbol{b})$ when fed with a one-hot encoded target vector to produce a distribution (generally non-uniform in nature) over components, from which one can draw one for sampling. As we can see in Table 1, the obtained accuracies for the linear classifier are not excessive, since GMM responsibilities are not optimized for discrimination. Figure 2 shows both unconditional and conditional sampling results for GMMs trained for 100 epochs on MNIST, FashionMNIST and Fruits-360. Conditional sampling is used to generated samples of class 5, and we can visually confirm that virtually all produced samples belong to this class. All reported results are accuracies for single runs, measured on the test test of the respective benchmarks.

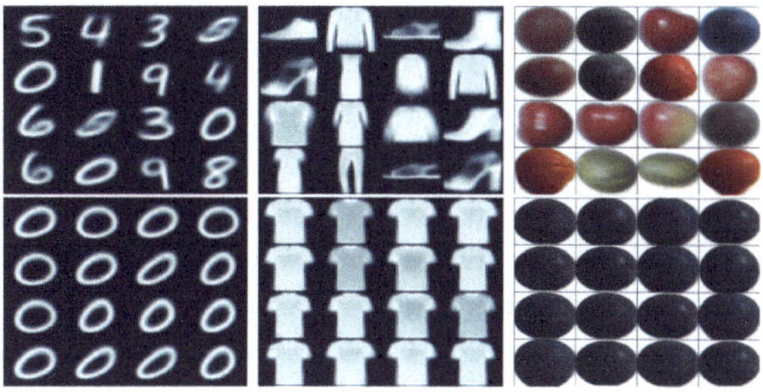

Fig. 2. Samples generated by GMMs with diagonal covariance matrix after training on MNIST (left column), FashionMNIST (middle column) and Fruits-360 (right column). In the upper row, sampling is unconditional, whereas we conditionally sample from class 5 in the lower row.

3.3 Sampling with MFA Instances

Since MFA instances can be considered to be GMMs whose covariance matrix interpolates between diagonal and full form, we can expect more diverse sampling results from MFA instances than from GMMs with diagonal covariance matrices. This is significant since, for high-dimensional data, full covariance matrices would require excessive memory.

We train an MFA model with a latent dimension of $l = 4$ on all datasets described in Sect. 3 for 300 epochs using $K = 100$ components. The samples generated by selected components of this model are shown in Fig. 3. To show that

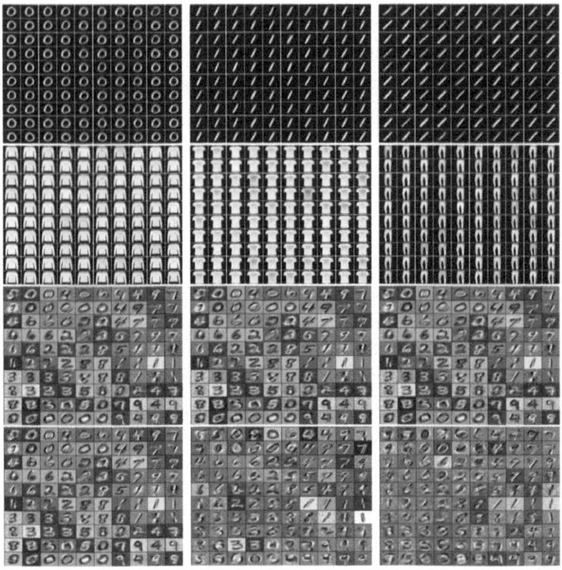

Fig. 3. Upper two rows: Sampling results for MFA on MNIST (first row) and FashionMNIST (second row). Each image shows 100 samples produced from the same component. Lower two rows: the first three latent directions for MFA on MNIST (3^{rd}row) and FashionMNIST (4th row).

the variability is caused by the latent factors and not by different components being selected for sampling, we always draw samples from a single component. We indeed observe that samples are quite diverse in terms of slant and stroke even when coming only from one component. This is especially true when comparing MFA-generated samples to those obtained from diagonal-CM GMMs (see previous section).

The latent directions are shown in Fig. 3. Latent directions are the columns of the matrices Γ_k, corresponding to (generalized, since the Γ_k are not full rank) principal directions. Latent direction i indicates the variations of sampled data that occur if we move along one axis i in the latent space and are useful for visualizing the sample variability captured by the MFA model.

3.4 Outlier Detection Experiments with MFA

For these experiments, we train a single MFA on class 5 of the MNIST and FashionMNIST datasets (Fruits was omitted since it is a very easy classification benchmark), and impose a threshold on the log-probability of the best-matching component in order for a given sample to be classified as an inlier. By varying this threshold, we obtain a receiver-operator characteristic (ROC) that indicates the quality of the outlier detector. As we can see in Fig. 4, outlier detection is clearly something that GMMs can perform well. We observe that increasing the number of latent factors increase the quality of outlier detection. Increasing the number of components(not shown) has a similar effect which is however not as pronounced.

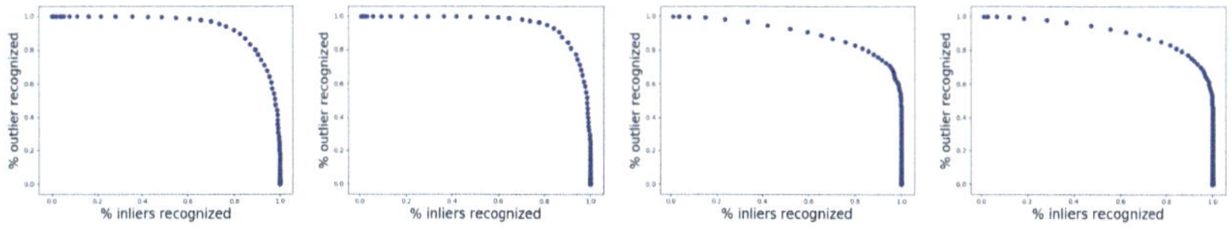

Fig. 4. Outlier detection quality. Shown, from left to right, are ROCs for MNIST (4 latent factors), MNIST (25 latent factors), FashionMNIST (4 latent factors) and FashionMNIST (25 latent factors). Inlier class is always class 5. The area under the curve is a performance indicator and follows a strict the-larger-the-better principle.

3.5 MFA: Generative Classification

Generative classification is in essence outlier detection with multiple outlier detection models $\mathcal{M}_i$, each model being trained on class i only. For testing, each model produces a score for the test sample, and the model with the highest score defines the predicted class for that sample: $p(c|\boldsymbol{x}) = \text{argmax}_i \mathcal{M}_i(\boldsymbol{x})$. If the outlier detection models are of sufficient quality, this scheme is not more computationally demanding than training a conventional DNN classifier, because although there are now several models, each model only gets trained on a small part of the data. We again found that this works best with MFA and a rather high latent dimensionality, e.g., $l = 25$. For reference, we also train conventional GMMs for this problems and compare the results (see Table 2. All reported results are accuracies for single runs, measured on the test of the respective benchmarks. It is important to note that generative classification is actually supervised learning: although the labels are not used while training the individual outlier detectors, they *are* used for assigning data samples *to* outlier detectors for training.

Table 2. Results for generative classification using MFA and GMM outlier detection models on various datasets.

dataset	Fruits-360	MNIST	FashionMNIST
GMM	91.0	85.3	81.2
MFA, $l = 4$	92.6	85.9	83.3
MFA, $l = 25$	98.5	95.7	92.3

4 Discussion and Outlook

In this article, we argue that SGD-trained Gaussian Mixture Models can be used as a drop-in replacement for SOMs in data visualization tasks that SOMS are typically used for. In particular, training GMMs by SGD enables them to process large amount of high-dimensional data which is required for data visualization and other typical data science tasks. We showed that SOMs can be considered an approximation to GMMs in a particular, simplifying case, and that SOMs do have a probabilistic interpretation. However, we argue that GMMs are much better suited than SOMs for performing typical data science functions such as outlier detection, sampling and generative classification because they offer a much greater flexibility in terms of, e.g., parameterizing the covariance matrix.

References

1. Gepperth, A.: Large-scale gradient-based training of mixture of factor analyzers. In: International Joint Conference on Neural Networks (IJCNN) (2022)
2. Gepperth, A., Pfülb, B.: Gradient-based training of Gaussian mixture models for high-dimensional streaming data. Neural Process. Lett. (2021, accepted)
3. Heskes, T.: Energy functions for self-organizing maps. In: Kohonen Maps, pp. 303–315. Elsevier (1999)
4. Heskes, T.: Self-organizing maps, vector quantization, and mixture modeling. IEEE Trans. Neural Netw. **12**(6), 1299–1305 (2001)
5. Kohonen, T.: The self-organizing map. Proc. IEEE **78**(9), 1464–1480 (1990)
6. LeCun, Y., Bottou, L., Bengio, Y., Haffner, P.: Gradient-based learning applied to document recognition. Proc. IEEE **86**(11), 2278–2324 (1998)
7. Mureşan, H., Oltean, M.: Fruit recognition from images using deep learning. Acta Universitatis Sapientiae Informatica **10**(1), 26–42 (2018). https://doi.org/10.2478/ausi-2018-0002
8. Richardson, E., Weiss, Y.: On GANs and GMMs. In: Advances in Neural Information Processing Systems (NeurIPS), pp. 5847–5858 (2018)
9. Verbeek, J.J., Vlassis, N., Kröse, B.J.: Self-organizing mixture models. Neurocomputing **63**, 99–123 (2005)
10. Xiao, H., Rasul, K., Vollgraf, R.: Fashion-MNIST: a Novel Image Dataset for Benchmarking Machine Learning Algorithms, pp. 1–6 (2017). http://arxiv.org/abs/1708.07747

A Self-Organizing UMAP for Clustering

Josh Taylor(✉) and Stella Offner

The University of Texas at Austin, Austin, TX, USA
joshtaylor@utexas.edu

Abstract. We introduce SOUMAP (Self-Organizing Uniform Manifold Approximation and Projection) as an alternative to regular Self-Organizing Map (SOM) learning intended to improve clustering of the resulting output space. Improvement is achieved by abandoning the SOM's rigid lattice structure in favor of a more expressive output topology afforded by Uniform Manifold Approximation and Projection (UMAP), which is incrementally learned in conjunction with SOUMAP prototypes. As in regular SOM learning, the Hebbian connection formed between input and output spaces results in a topologically trustworthy low-dimensional embedding amenable to clustering. Through controlled experiments we show that SOUMAP's more expressive "lattice" improves the quality of clusterings obtained from it.

Keywords: Self-Organizing Maps · Dimensionality Reduction · Clustering

1 Introduction

Dimensionality reduction (DR) techniques serve various purposes in data analysis and machine learning. As feature selectors, they are often used as a preprocessor to either reduce the computational burden or improve performance of subsequent learning stages. When used to project high-dimensional data to two or three dimensions, DR techniques allow visualization of high-d structures; such visualizations are often used to inform or validate data clusterings, and as scaffolding upon which summary statistics of high-d manifolds can be visualized in an organized manner. For success in these latter roles, it is crucial that the outputs of DR algorithms are fiducial representations of high-d structure.

Modern DR techniques such as t-Stochastic Neighbor Embedding (t-SNE, [8]) and Uniform Manifold Approximation and Projection (UMAP, [10]) have become commonplace tools for data summarization across research disciplines, with heavy use in the life and physical sciences. Despite recent work shedding light on how t-SNE and UMAP produce such expressive embeddings [1,2] confusion persists about about how faithfully either preserve underlying data structure. This confusion arises partly from potential user (mis-)parameterization, but also from optimization of their cost functions [2,3].

Authors acknowledge partial support from NSF AAG 2107942 and seed funding from NSF's Institute for Foundations in Machine Learning.

T. Villmann et al. (Eds.): WSOM+2024, LNNS 1087, pp. 63–73, 2024.
https://doi.org/10.1007/978-3-031-67159-3_8

On the other end of the (dis-)trust spectrum is Kohonen's Self-Organizing Map (SOM), whose non-linear projections are constrained to a low-dimensional (usually 2- or 3-d) lattice. While this constraint typically results in a less expressive embedding of high-d manifolds, SOM embeddings are assumed to be (approximately) topology-preserving due to both their robustness to (mis-) parameterization and their explicit and simple Hebbian learning link between low- and high-d spaces [6]. This link establishes a feedback mechanism between learned representations in input and output spaces, bolstering our trust in both at termination of learning. Further, associated visualizations [12] readily convey where a SOM is *not* topology preserving, which provides insurance against faulty inferences from its learning.

This work is motivated by a desire to combine the visual expressiveness of t-SNE or UMAP DR with the trustworthiness and inferential confidence offered by the SOM. We achieve this by replacing the rigid SOM lattice with a more nimble low-d topology obtained via UMAP during iterative self-organizing learning. Thus, our contributions can be thought of as either *adding* a competitive learning phase to UMAP, or *removing* the dependence of the SOM on its lattice; both are correct. As in Kohonen's mapping, and in contrast to similarly motivated work [5], SOUMAP's learning rule inextricably marries its prototype formation to the topology of its (prevailing) output space which bolsters confidence that the latter accurately represents the former. In the following sections we outline the methodology for a Self-Organizing UMAP (SOUMAP) (Sect. 3) and highlight experiments showing its utility (Sect. 5).

We stress that SOUMAP is a flexible framework marrying competitive Hebbian learning with alternative output space topologies, which could arise from any DR technique. Here, we have chosen UMAP due to its speed but could just as easily have prescribed a Self-Organizing {t-SNE, Autoencoder, ISOMAP, ...} following the same logic presented here.

2 Background

2.1 The Self-Organizing Map

A SOM is an unsupervised neural network that mimics how the cortex of the human brain summarizes and organizes information. SOMs consists of a rigid (usually 2-d) lattice $\mathscr{L} = \{l_i\}_{i=1}^M$ of M artificial neurons, each of which is connected to its input space by a $d-$dimensional weight vector known as a *prototype*. In this work we denote prototypes by $W = \{w_i \in \mathbb{R}^d\}_{i=1}^M$. The SOM learns a sample $X = \{x_s \in \mathbb{R}^d\}_{s=1}^N$ of size N, assumed to arise from manifold $\mathscr{M}$ with density $f(x)$, by simultaneously (1) positioning its $M << N$ prototypes in $\mathbb{R}^d$ to estimate f and (2) organizing the prototypes' associated neurons on $\mathscr{L}$ to represent the structure of $\mathscr{M}$. Thus, the SOM performs vector quantization and nonlinear DR. SOM-based clusterings utilize both of these learned results to partition data.

The topology-preserving SOM mapping $\mathscr{M} \to \mathscr{L}$ is formed via an iterative two-step learning rule relying on competition (among the prototypes to represent each datum) and cooperation (among lattice neurons as prototypes are updated). The latter is important to this project so we reiterate: the SOM lattice $\mathcal{L}$ influences its prototype formation during learning. This influence makes the SOM robust to spurious network initialization and bestows the topology-preserving properties upon the learned mapping.

To be more explicit we define $BMU1(s) = \arg\min_i \delta(x_s, w_i)$ as the best matching unit (i.e., the index of the prototype of minimum distance) of a particular observations x_s. We have let $\delta(\cdot,\cdot)$ denote an arbitrary distance, although it is typically prescribed as Euclidean. At the start of learning, prototypes w_i and their associated lattice positions l_i are initialized (either randomly, or via some weakly-organizined manner such as a Principal Components Analysis of the underlying data, [6]). At each iteration t, batch SOM learning (which we use in this work) updates the prototypes as a weighted average of all input data,

$$w_i(t) = \frac{\sum_{s=1}^{N} \eta(i, BMU1(s), t) \star x_s}{\sum_{s=1}^{N} \eta(i, BMU1(s), t)}, \tag{1}$$

where the weighting is determined by the neighborhood function η, defined on $\mathcal{L}$. η is typically a Gaussian smoothed by a time-annealed $\sigma(t)$:

$$\eta(i, j, t) = \exp\left(-\delta(l_i, l_j)^2 / 2\sigma(t)^2\right), \tag{2}$$

where, here, $\delta(l_i, l_j)$ is usually a geodesic distance on $\mathcal{L}$. We denote geodesics by $\Delta(\cdot,\cdot)$ in what follows. Annealing $\sigma(t)$ from some initially large value to a terminal small value $\sigma(T) \approx 1$ ensures organization of $\mathcal{L}$ with respect to W.

2.2 UMAP

UMAP [10] learns an embedding $Y = \{y_i \in \mathbb{R}^{d'}\}_{i=1}^{N}$ of arbitrary high-d points $P = \{p_i \in \mathbb{R}^d\}_{i=1}^{N}$, where $d' << d$ is user-specified at the onset of learning. Y arises via stochastic gradient descent of the UMAP cost function

$$C_U = -\sum_{ij} \left[\underbrace{\mu_{ij} \log(\phi_{ij})}_{\text{attractive}} + \underbrace{(1 - \mu_{ij}) \log(1 - \phi_{ij})}_{\text{repulsive}} \right]. \tag{3}$$

The μ_{ij} terms are similarities $\in [0, 1]$ between high-d points p_i and p_j:

$$\mu_{ij} = (\mu_{j|i} + \mu_{i|j}) - \mu_{j|i}\mu_{i|j}, \quad \mu_{j|i} = e^{-\beta(\delta(p_i, p_j) - \rho_i)}, \quad \rho_i = \min_{j' \neq i} \delta(p_i, p_{j'})$$

where $\delta(\cdot,\cdot)$ is usually Euclidean. ρ_i is the nearest neighbor distance from p_i, so that $\mu_{j|i}$ is a smoothed similarity between p_i and all other points such that $\max \mu_{j|i} = 1$. β is a smoothing parameter found via binary search such that $\sum_j \mu_{j|i} = \log_2(k_U)$. $\mathbf{k_U}$ is the most important user-supplied parameter to UMAP controlling the number of nearest-neighbors UMAP attempts to preserve in its embedding. μ_{ij} merely symmetrizes the $\mu_{j|i}$.

Likewise, the ϕ_{ij} terms of (3) are similarities of the associated points y_i and y_j in the low-d embedding, given by $\phi_{ij} = (1 + a \star \delta(y_i, y_j)^{2b})^{-1}$. ϕ is a generalized Cauchy probability density, where tail thickness is controlled by a and b. Minimization of UMAP's cost proceeds by stochastic gradient descent for a specified number of epochs T_U.

3 SOUMAP Methodology

SOUMAP is a hybrid, iterative paradigm linking competitive and cooperative SOM learning (Sect. 2.1) with a UMAP embedding (Sect. 2.2) of the prevailing SOM prototypes at each iteration.

SOUMAP initializes a regular SOM lattice and, after T_ϵ iterations of SOM learning performs a small number (T_U) of UMAP iterations on the prevailing values of W, replacing the neuron locations l_i with the UMAP result y_i. Prototype updates proceed as normal, except the distance δ in the neighborhood activation function η (2) is now computed with respect to the topology of Y. For this purpose, we compute the Delaunay graph $\mathscr{D}_Y$ [4] of embedded points Y, and measure the geodesic $\Delta(\cdot,\cdot)$ along $\mathscr{D}_Y$, instead of along $\mathscr{L}$. From here on, every SOUMAP iteration t involves a prototype update, and every T_ϵ iterations another UMAP update is performed to modify $Y(t)$, using $Y(t-T_\epsilon)$ as UMAP's initial coordinates.

By allowing the neurons to abandon their initial lattice positions and slowly drift toward something deemed more optimal (by UMAP), SOUMAP has the capacity to produce a terminal configuration that is more expressive than $\mathscr{L}$, and more topology-preserving than UMAP would produce on its own, if the terminal SOM prototypes were embedded in a subsequent (separate) stage post-learning. We present the overall SOUMAP algorithm pseudocode in Algorithm 1 and address several of its more exotic steps (compared to regular SOM learning) in the remainder of this section.

During early SOM learning, larger values of $\sigma(t)$ tend to push many SOM prototypes W to very similar values (major density peaks) for a short period of time; this behavior is rectified as $\sigma(t)$ is annealed, which allows the prototypes W to represent more nuanced regions of the underlying data density. This is perfectly normal SOM behavior, where the lattice helps the prototypes unwind to their ultimate refined positions.

However, UMAP is particularly sensitive to the high degree of correlation among these infant prototypes and responds by embedding many prototypes to nearly identical Y at early stages. Because we use $Y(t - T_\epsilon)$ as UMAP's initialization for the UMAP update at iteration t, this imparts lasting effects on Y. Refraining from any UMAP updates until the SOM enters its convergence phase would mitigate this, but would also prevent some of the clustering benefits of SOUMAP discussed in Sect. 5. Instead, we have adopted a strategy where, during early learning, UMAP updates are performed on the prevailing prototypes at iteration t *in conjunction* with a sub-sample $S(t)$ of X of size $\gamma(t)N$ which decreases with t. That is, UMAP is applied to $W \cup S(t)$, and the prototype

Algorithm 1. SOUMAP

Inputs: X (data), $\mathscr{L}_x, \mathscr{L}y$ (lattice width and height), T_ϵ (UMAP update frequency), T (max. iterations), σ_0 (starting width of SOM neighborhood), T_U (number of UMAP epochs performed during incremental UMAP updates to Y)

Initialization: of SOM lattice $\mathscr{L}$, geodesic distances Δ along $\mathscr{L}$, and prototypes W

Learning:
$t \leftarrow 1$
while $t \leq T$ **do**
  $\sigma(t) \leftarrow$ anneal neighborhood radius
  $W \leftarrow$ SOM prototype update, using η with Δ and $\sigma(t)$
  if $t \mod T_\epsilon = 0$ **then**
    $\gamma(t) \leftarrow$ anneal sub-sampling rate, sample $S(t)$ from X
    $Y \leftarrow$ UMAP projection (T_U epochs) of $W \cup S(t)$
    $\mathscr{D}_Y \leftarrow$ Update Delaunay graph of Y
    $\Delta \leftarrow$ Update geodesic distances along $\mathscr{D}_Y$
  $t \leftarrow t + 1$

embeddings are extracted from this (joint) UMAP output. $\gamma(t) = \sigma^2(t)/|\mathscr{L}|$ is the sub-sampling rate at iteration t, where $|\mathscr{L}|$ is the cardinality of $\mathscr{L}$. Over time, $|S(t)|$ decreases, ensuring the embedded prototypes Y reflect only their high-d counterparts W at termination. We believe this sub-sampling strategy predominately necessary when UMAP is used as the DR technique to produce Y; UMAP's sensitivity appears to stem from the oddities of its optimization routine [3], as we do not observe similar behaviour when, e.g., t-SNE is used instead.

4 Experiment Descriptions

4.1 Datasets, Parameterization, and Design

To assess SOUMAP performance for both of its tasks (prototype learning and DR) we have conducted a series of experiments on samples of size $N = 20,000$ from the following datasets: **worms64**[1] is a synthetic dataset of dimension $d = 64$ containing 25 clusters of "worm-like" shapes; **MNIST**, **Fashion-MNIST**[2], and **Kazushiji-MNIST**[3] are commonly used databases of 28×28 pixel ($d = 784$) grayscale images of handwritten digits 0-9, 10 different articles of clothing, and 10 different Japanese Hiragana characters, respectively.

For each dataset above, we perform regular (lattice-based) batch SOM learning ($M \approx 2000$ prototypes for worms64, $M \approx 3600$ for MNIST and its relatives)and compare it to SOUMAP learning for the same number of total epochs

[1] https://cs.uef.fi/sipu/datasets/.
[2] https://github.com/zalandoresearch/fashion-mnist.
[3] https://github.com/rois-codh/kmnist.

$T = 50$. For SOUMAP we allow UMAP updates of $T_U = 100$ epochs for every $T_\epsilon = 5$ of its iterations. The initial SOM geodesic radius σ_0 was set to 25% of the hexagonal lattice width and annealed exponentially to $\sigma(T = 50) = 1$. Thus, both the SOM and SOUMAP receive 50 batch updates to their prototypes; SOUMAP receives a total of 1,000 ($\frac{T=50}{T_\epsilon=5} \times (T_U = 100)$) epochs of UMAP learning of the evolving prototypes. For now, we have kept all UMAP parameters (in particular, number of neighbors $k_U = 15$) at their defaults. The increased computational time for UMAP is negligble for these data; worst cases (MNIST and its relatives, where $M \approx 3600$ 784-d prototypes) require less than 10 additional seconds for all 1,000 UMAP training epochs. To investigate a simpler analog to SOUMAP we also compute prototypes via K-means (using the same M discussed above for SOM & SOUMAP), and embed them with default UMAP. We denote this two-stage approach, where prototype learning and embedding are sequentially separate processes, by KM+U. All three prototype generation methods and their associated embeddings are compared in Sect. 5.

4.2 Quality Measures

We assess the performance of SOM and SOUMAP learning by measuring the topology preservation and clustering amenability of each method.

Topology preservation measures (TPMs) assess how well high-d neighborhoods are preserved when represented in a low-d embedding. Numerous TPMs exist, varying mostly in their definition of "neighborhood." We have selected two for consideration: the k-Ary score, and the Weighted Average Folding Length. The k-Ary score [7] measures the proportion of a high-d k-nearest neighborhood around each prototype w_i that is preserved after embedding w_i as y_i in low-d, averaged over all $i \in \{1, \ldots, M\}$ prototypes. As this K-ary measure ([7, equation 15]) yields a performance curve over $k \in \{1, \ldots, M-1\}$, we report the area under such curve (AUC), normalized by its theoretical maximum (M-2), for comparison across datasets. k-Ary is in the range $[0, 1]$, with 1 indicating perfect preservation of high-d neighborhoods of all sizes k.

The Weighted Average Folding Length (WAFL) [13] is a more local measure of TP *with respect to* the data manifold. WAFL is based on the CADJ graph with prototypes as vertices [12]. CADJ is a weighted version of the Induced Delaunay Graph [9], which estimates topological adjacencies within the manifold $\mathcal{M}$ from which X was sampled. Such adjacencies may be different, e.g., when two prototypes are nearest Euclidean neighbors, but no data X exists between them. Thus, CADJ is tuned to highlight seams in the learned manifold caused by its disconnection at cluster boundaries. CADJ edge weights are given by

$$CADJ_{ij} = \sum_s I(BMU1(s) = i \wedge BMU2(s) = j) \tag{4}$$

where $I(\cdot)$ is the indicator function, and $BMU1(s)$ and $BMU2(s)$ return the indices of the first and second best matching units (nearest, and second-nearest prototypes) to datum x_s, respectively. The WAFL, defined as

$$WAFL = \frac{1}{N}\sum_{ij} CADJ_{ij} \star \Delta_{ij}, \tag{5}$$

reports the weighted average geodesic neighborhood radius along the Delaunay graph $\mathscr{D}_Y$ of embedded prototypes required to include all high-d CADJ neighbors. Thus, it measures the degree to which each CADJ edge is preserved as a nearest geodesic neighbor in embedded space. An ideal value $WAFL = 1$ indicates all CADJ neighbors are preserved, while any $WAFL > 1$ grades the mis-preservation accordingly.

Clustering assessments of both the learned SOUMAP prototypes, and their projection Y, can help determine the suitability for SOUMAP as a clustering tool. To this end, we compute an average of clustering quality metrics (the Adjusted Rand Index – ARI, and Adjusted Mutual Information – AMI) using the known data labels as ground truth; as each is more applicable in certain conditions [11] we have averaged them into a single measure, $cl = (AMI + ARI)/2$. To cluster learned W or Y we employ hierarchical agglomerative clustering with single (HACS), complete (HACC), average (HACA), and Ward (HACW) linkages, compute cl for each, and accept the maximum resulting cl value as the reported metric for each dataset. We cluster both W and Y in this manner, producing quality measures for both of SOM/SOUMAP's prototypes and resulting embeddings.

5 Results

Visual results from our experiments with each dataset are presented in Fig. 1, where the final SOM embedding is shown at left, and the final SOUMAP embedding shown at right. In each panel, neurons are colored by their corresponding class label, which was assigned by plurality vote of the class labels of data in each neuron's receptive field (RF), which is defined as the set of $\{x_s\}$ mapped to each neuron i. Each neuron's plotted size is scaled to the size of its receptive field; dead neurons (those representing no data) have been removed. Additionally, the thin gray lines connecting neurons depict the edges of the CADJ graph (4). This visualization technique, which is an un-weighted analog of the CONNvis visualization [12], helps identify topology violations; maps with better topology preservation have overall shorter gray edges with minimal edge crossings.

Visual inspection of Fig. 1 reveals strong organization for both the SOM and SOUMAP. However, the final SOUMAP embeddings (right column) contain more tightly packed clusters and, hence, more separation between clusters, contributing to a more expressive representation of overall cluster structure. This is the hallmark characteristic of UMAP which motivated SOUMAP in the first place. SOUMAP's early abandonment of the lattice has visually appealing benefits, particularly for the more complicated data considered in these experiments. For example, the SOM has signaled, to the best of its ability, a boundary between the blue and orange clusters of FMNIST (representing images of boots and handbags, respectively); SOUMAP's delineation of this boundary clearly

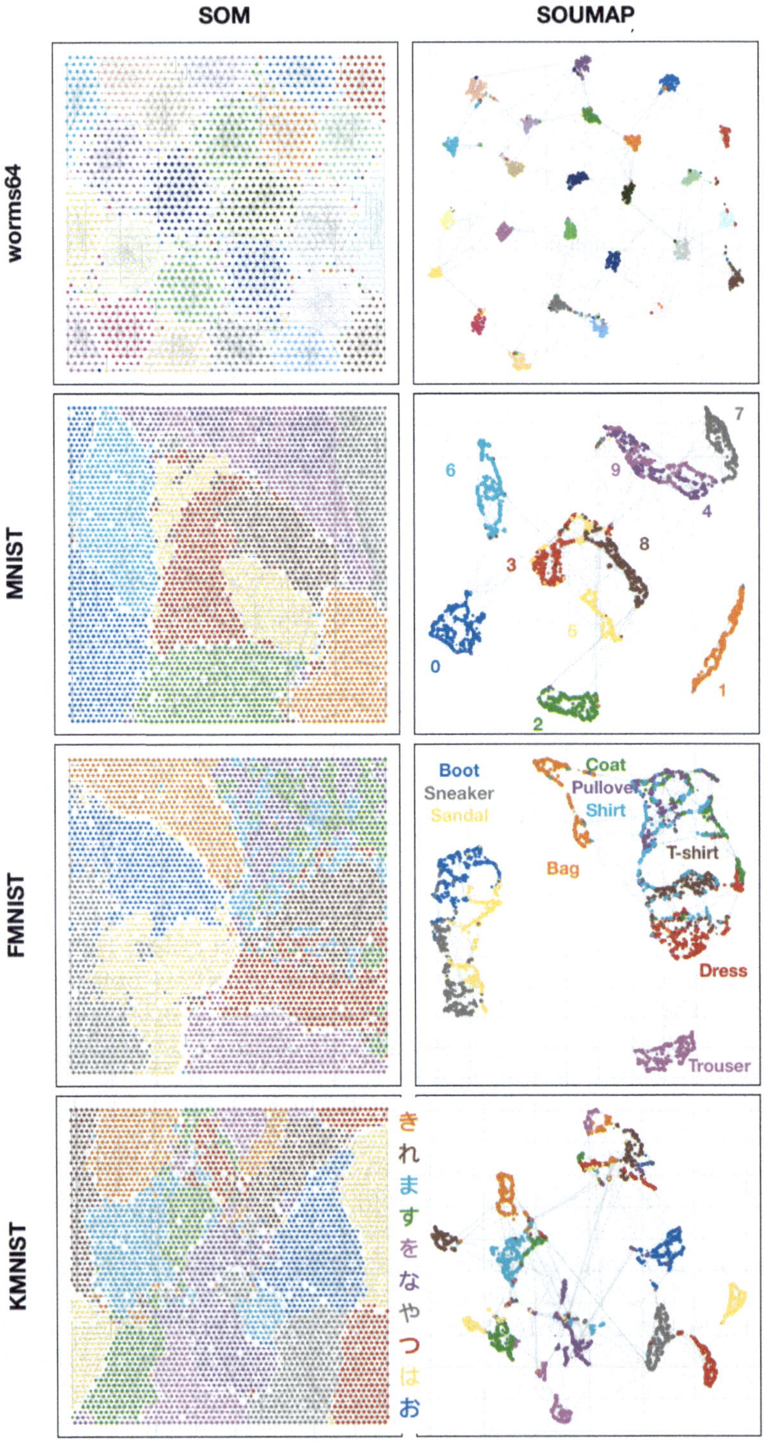

Fig. 1. Learned prototype embeddings of our experimental data from Sect. 4.

Table 1. Quantitative Performance Measures, as defined in Sect. 4.2. Directional arrows indicate better performance according to each measure.

			worms64	MNIST	FMNIST	KMNIST
Topology	k-Ary ↑	SOM	0.56	0.60	0.65	0.61
		SOUMAP	0.54	0.54	0.58	0.55
		KM+U	0.21	0.42	0.44	0.41
	WAFL ↓	SOM	1.37	1.44	1.24	1.34
		SOUMAP	1.28	1.23	1.16	1.17
		KM+U	5.11	3.02	2.83	2.88
Clustering	W ↑	SOM	0.70	0.54	0.46	0.44
		SOUMAP	0.73	0.62	0.51	0.45
		KM+U	0.00	0.17	0.18	0.13
	Y ↑	SOM	0.52	0.41	0.42	0.34
		SOUMAP	0.73	0.75	0.52	0.51
		KM+U	0.48	0.77	0.49	0.48

separates bags not only from boots but also from the mixed coat/pullover/shirt super class. More pronounced separation is visible for SOUMAP's KMNIST embedding; the SOM only weakly signals separation between the orange/brown, green/pink, and gray/red character classes, but SOUMAP shows marked delineation for these characters.

From Table 1 we observe that, of the two TPMs considered here, k-Ary and WAFL report divergent trends: the SOM has a consistently higher (better) k-Ary score, but also a consistently higher (worse) WAFL score, than SOUMAP. Our UMAP parameterization is the cause of this behavior, as UMAP was only instructed to attempt preservation of the $k_U = 15$ nearest high-d neighbors; it is doing so, but at the expense of higher-order neighborhoods. This is a known trade-off between global and local views of data in DR techniques. However, as the CADJ graph has proven a useful tool for SOM clustering, we believe fiducial representation of CADJ neighborhoods more important than representation of neighborhoods of all sizes, which is what k-Ary is designed to measure.

Turning to the clustering quality metric of Table 1, we see SOUMAP's strength is producing embeddings Y that are more amenable to clustering than the regular SOM. This agrees with our visual assessments above which found SOUMAP more expressive in general. Interestingly, there is also some small improvement when clustering SOUMAP's prototypes W directly; we assume this is a result of SOUMAP's output topology becoming tightly organized early in the learning process. If such embedded structures are "real" (meaning they faithfully represent analogous structure in high-d), it should be expected that the learning reinforcement provided by the neighborhood function η would help refine the high-d prototypes and bolster prototype clusterings.

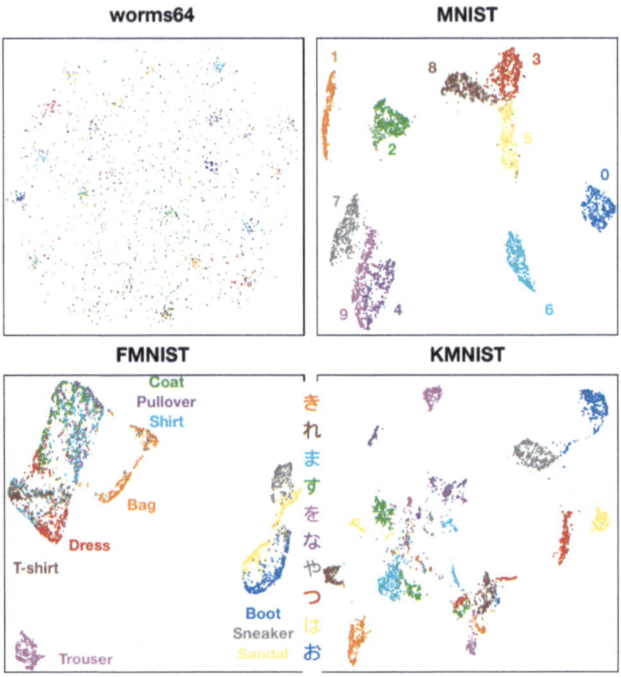

Fig. 2. K-means + UMAP Embeddings of the data from Sect. 4.

The hybrid K-means + UMAP baseline strategy produces seemingly coherent embeddings of our data (Fig. 2) but underperforms both the SOM and SOUMAP according to the quantitative measures of Table 1. While the embeddings resulting from KM+U possess clusterability similar to SOUMAP, the K-means prototypes are not nearly as amenable to clean clustering. Additionally, the low-d mapping of the two-stage KM+U procedure suffers from large degradation in topology preservation. Combined, these results indicate there is much to be gained from integrating the prototype learning and DR tasks, vs. performing them sequentially.

6 Conclusions and Future Work

We have presented SOUMAP, a hybrid prototype learning scheme to harness all of the benefits of SOM-based vector quantization (sample size and noise reduction) in conjunction with the expressive low-d visualizations of data provided by UMAP. Our experiments show that SOUMAP is capable of both preserving the high quality of SOM-based VQ while simultaneously providing visually appealing embeddings that are more amenable to cluster inference.

With the prototype learning rule established, the next stage in SOUMAP development is a deeper analysis into the parameterization of its UMAP update stage. Ultimately we would prefer SOUMAP to be completely self-parameterized and, with so many parameters at play, our UMAP stage would likely benefit from parameter tuning. [13] outlines an automated procedure for specifying t-SNE's perplexity parameter based on an analysis of the CADJ graph (4); as UMAP shares a similar parameter (number of neighbors, denoted k_U here), we believe a related analysis may partly contribute to a self-governed SOUMAP updating process. Because UMAP is more complex than t-SNE (i.e., has more parameters, and a wildly different optimization procedure), expanding upon [13] in a UMAP-specific context is a necessary next step.

References

1. Böhm, J.N., Berens, P., Kobak, D.: Attraction-repulsion spectrum in neighbor embeddings. J. Mach. Learn. Res. **23**(95), 1–32 (2022)
2. Damrich, S., Böhm, N., Hamprecht, F.A., Kobak, D.: From *t*-SNE to UMAP with contrastive learning. In: The Eleventh International Conference on Learning Representations (2022)
3. Damrich, S., Hamprecht, F.A.: On UMAP's true loss function. Adv. Neural. Inf. Process. Syst. **34**, 5798–5809 (2021)
4. Fortune, S.: Voronoi diagrams and delaunay triangulations. In: Computing in Euclidean Geometry, pp. 225–265 (1995)
5. Fritzke, B.: Growing cell structures-a self-organizing network for unsupervised and supervised learning. Neural Netw. **7**(9), 1441–1460 (1994)
6. Kohonen, T.: Self-Organizing Maps, vol. 30. Springer, Heidelberg (2012). https://doi.org/10.1007/978-3-642-56927-2
7. Lee, J.A., Peluffo-Ordóñez, D.H., Verleysen, M.: Multi-scale similarities in stochastic neighbour embedding: reducing dimensionality while preserving both local and global structure. Neurocomputing **169**, 246–261 (2015)
8. Van der Maaten, L., Hinton, G.: Visualizing data using t-SNE. J. Mach. Learn. Res. **9**(11) (2008)
9. Martinetz, T., Schulten, K.: Topology representing networks. Neural Netw. **7**(3), 507–522 (1994)
10. McInnes, L., Healy, J., Melville, J.: UMAP: uniform manifold approximation and projection for dimension reduction. arXiv preprint arXiv:1802.03426 (2018)
11. Romano, S., Vinh, N.X., Bailey, J., Verspoor, K.: Adjusting for chance clustering comparison measures. J. Mach. Learn. Res. **17**(1), 4635–4666 (2016)
12. Tasdemir, K., Merényi, E.: Exploiting data topology in visualization and clustering of self-organizing maps. IEEE Trans. Neural Networks **20**(4), 549–562 (2009)
13. Taylor, J., Merényi, E.: Automating t-SNE parameterization with prototype-based learning of manifold connectivity. Neurocomputing **507**, 441–452 (2022)

Knowledge Integration in Vector Quantization Models and Corresponding Structured Covariance Estimation

M. Kaden(✉), J. Voigt, K. S. Bohnsack, M. Lange-Geisler, and Thomas Villmann

Saxon Institute for Computational Intelligence and Machine Learning (SICIM), Mittweida University of Applied Science, Mittweida, Germany
{kaden1,voigt5,villmann}@hs-mittweida.de

Abstract. We propose a knowledge-informed variant for learning covariance-dependent data representations using unsupervised vector quantization. In particular, we consider linear data mappings included in vector quantization models, such as c-means++, neural gas, or self-organizing maps, to achieve representations in a lower-dimensional data space. To this end, we show how additional data knowledge can be integrated into the models. The additional data structure information is used to generate an appropriate data mapping depending on the corresponding structured data covariances.

Keywords: Knowledge-informed covariance estimation · unsupervised vector quantization · neural gas · self-organizing maps · principal component analysis

1 Introduction

Efficient models for vector quantization are still of great interest in areas such as astrophysics, bioinformatics, media, and others. The goal of these algorithms is to generate a set of codebook vectors, often called prototypes, that act as representatives of the original data. These prototypes should retain as much information as possible from the original data. Often, Principal Component Analysis (PCA) is used for data preprocessing and complexity reduction [8].

The representation of the data is usually realized in these models by the nearest prototype principle (NPP), i.e. a data point is assigned to the prototype which has the minimum deviation for a pre-defined dissimilarity measure.

Remark 1. *It should be noted that one could alternatively investigate the NPP in terms of maximum similarity [13], which would be more intuitive in terms of cognitive science where similarity between objects/subjects is considered in terms of the number of shared properties [15]. However, this approach has the difficulty that optimal similarity cannot generally be identified with an upper bound on the similarity value, whereas NPP based on dissimilarities has a zero lower bound for the optimal solution. Therefore, we focus on this case here.*

T. Villmann et al. (Eds.): WSOM+2024, LNNS 1087, pp. 74–85, 2024.
https://doi.org/10.1007/978-3-031-67159-3_9

Depending on the task and the given data, adequate dissimilarity measures should be chosen. Yet, a common choice is the (squared) Euclidean distance for vectorial data. The underlying assumption for this choice is that the variance of the data contributes to the information content of the data. Another important information criterion that must be explicitly or implicitly taken into account by vector quantization models is the data density or distribution, respectively. From a coding point of view, each prototype should represent approximately the same amount of data for optimal information preservation. This problem is known as the magnification problem [16,19]. When data density is considered to be the dominant feature for data representation, variance equalization should be applied as data pre-processing. In the case of vector quantizers based on Euclidean distance, the application of Mahalanobis distance for NPP instead of the Euclidean is equivalent to the variance equivalence pre-processing.

Probably the most prominent model for vector quantization is the c-means algorithm nowadays applied with intelligent initialization procedures like in c-means++ [1,19]. It assumes a set $\mathcal{P} = \{\mathbf{p}_1, \ldots, \mathbf{p}_c\} \subset \mathbb{R}^n$ of prototypes $\mathbf{p}_k$ to represent the data $\mathcal{X} = \{\mathbf{x} \in \mathbb{R}^n\}$. Thereby C is assumed to be the number of given clusters of the underlying data set. Optimization of $\mathcal{P}$ by c-means implicitly addresses also the magnification problem. More efficient vector quantization algorithms based on prototypes are the Self-Organizing Maps (SOM) and the Neural Gas model (NG) [11,12]. Both approaches also rely on the (squared) Euclidean metric and relate the distribution of the prototypes $\mathcal{P}$ to that of the data. Moreover, from a point of complexity reduction, PCA should be incorporated into this data processing as well.

Yet, a missing aspect for the application of these approaches is that no explicit knowledge about the data is taken into account. For example, if structural information about the data is available, a simple (co-)variance equalizing as pre-processing may be inappropriate because it does not reflect the dependencies/correlations which are valid between the data attributes. A consistent handling of the data should take care of this knowledge. In this contribution we propose an approach for vector quantization models to integrate structural information into the learning addressing particularly the variance problem. For this aim, we consider a parameterized Euclidean distance known from metric adaptation in classification learning [14], which has recently been adapted to knowledge-informed classification learning in gene expression data analysis [17]. We describe a possibility for incorporating structural knowledge into the vector quantization model and explain the resulting numerical behavior in comparison to classical approaches.

2 Vector Quantization by Self-Organizing Maps and Neural Gas

For SOM and NG, we assume a finite set of data $\mathcal{X} = \{\mathbf{x} \in \mathbb{R}^n\}$ with density $P_{\mathcal{X}}$ and a set $\mathcal{P} = \{\mathbf{p}_1, \ldots, \mathbf{p}_c\} \subset \mathbb{R}^n$ of prototypes. Further, the coding scheme of data vectors by means of $\mathcal{P}$ is given as the winner-takes-all (WTA) rule

$$s(\mathbf{x}) = \underset{j=1,\ldots,c}{\operatorname{argmin}} \{d(\mathbf{x}, \mathbf{p}_j)\} \tag{1}$$

for a given differentiable dissimilarity measure d, which is commonly the squared Euclidean distance $d(\mathbf{x}, \mathbf{p}_j) = (\mathbf{x} - \mathbf{p}_j)^2$. For SOM we further assume an external regular two-dimensional (neuron) grid A such that each prototype $\mathbf{p}_k$ is assigned to a grid location by the coordinate vector $\mathbf{r}_k \in A \subset \mathbb{R}^2$. Learning of the prototypes takes place by stochastic learning

$$\Delta p_j \propto h_\sigma(j, s(\mathbf{x})) \cdot (\mathbf{x} - \mathbf{p}_j) \tag{2}$$

for a randomly chosen data $\mathbf{x}$ according to $P_{\mathcal{X}}$. Here, $h(j, s(\mathbf{x}))$ is the so-called neighborhood function realizing a learning scheme based on the cooperative relations between neighbouring prototypes. SOM and NG differ regarding this function. In SOM the neighborhood learning is realized as

$$h_\sigma^{\text{SOM}}(j, s(\mathbf{x})) = \exp\left(-\frac{\left\|\mathbf{r}_j - \mathbf{r}_{s(\mathbf{x})}\right\|_A^2}{2\sigma^2}\right) \tag{3}$$

using an appropriate norm for the grid A. The quantity σ determines the spatial range of cooperativeness in A and is therefore denoted as neighborhood range. In NG, the neighborhood cooperativeness is determined according to

$$h_\sigma^{\text{NG}}(j, s(\mathbf{x})) = \exp\left(-\frac{\operatorname{rk}(j, s(\mathbf{x}))}{2\sigma^2}\right) \tag{4}$$

where the winning-rank function

$$\operatorname{rk}(j, s(\mathbf{x})) = \sum_{k=1}^{c} H(d(\mathbf{x}, \mathbf{p}_k) - d(\mathbf{x}, \mathbf{p}_j))$$

yields the winning rank of the prototype $\mathbf{p}_k$ according to the dissimilarity d.

$$H(z) = \begin{cases} 1 & z < 0 \\ 0 & \text{else} \end{cases}$$

is the Heaviside function. Further note that $\operatorname{rk}(s(\mathbf{x}), s(\mathbf{x})) = 0$ is valid. The parameter $\sigma > 0$ is the neighborhood range, yet in $\mathbb{R}^n$.

We can define the function

$$f(\mathbf{x}, \sigma) = \sum_{j=1}^{c} h_\sigma(j, s(\mathbf{x})) \cdot d(\mathbf{x}, \mathbf{p}_j)$$

with the expectation

$$E_{\mathcal{X}}[f(\mathbf{x}, \sigma)] = \int_{\mathcal{X}} f(\mathbf{x}, \sigma) \cdot P_{\mathcal{X}}(\mathbf{x})\, d\mathbf{x} \tag{5}$$

in dependence on the density $P_{\mathcal{X}}$. It turns out that

$$E(\mathcal{X}, \mathcal{P}, \sigma) = E_{\mathcal{X}}[f(\mathbf{x}, \sigma)] \tag{6}$$

is minimized applying the learning scheme (2) in case of NG using the squared Euclidean distance taken as stochastic gradient descent learning (SGDL) with

$$E(\mathbf{x}, \mathcal{P}, \sigma) = f(\mathbf{x}, \sigma) \cdot P_{\mathcal{X}}(\mathbf{x}) \tag{7}$$

as the local errors. For SOM, this statement is not valid [4]. However, if neighborhood cooperativeness is incorporated into the WTA-rule by

$$s(\mathbf{x}) = \operatorname{argmin}_{j=1,\ldots,c} \left\{ \sum_{k=1}^{c} h_{\sigma}^{\mathrm{SOM}}(j,k)\, d(\mathbf{x}, \mathbf{p}_k) \right\} \tag{8}$$

then $E_{\mathcal{X}}[f(\mathbf{x}, \sigma)]$ serves as a cost function minimized by the SOM-learning [7].

2.1 Adaptive Covariance Learning in SOM/NG

As already mentioned in the introduction, frequently all variances are equalized as a pre-processing step in machine learning. In NG and SOM, we can alternatively make use of the squared Mahalanobis distance

$$d_{\mathbf{C}}(\mathbf{x}, \mathbf{p}_j) = (\mathbf{x} - \mathbf{p}_j)^T \cdot \mathbf{C}^{-1} \cdot (\mathbf{x} - \mathbf{p}_j) \tag{9}$$

for the WTA-rules (1) or (8) instead of the standard squared Euclidean distance, where $\mathbf{C} = E_{\mathcal{X}}\left[\mathbf{x} \cdot \mathbf{x}^T\right]$ is the covariance matrix of the data, that is assumed to be regular.

As discussed in [10], one can estimate the covariance matrix during NG/SOM training using the dissimilarity measure

$$d_{\boldsymbol{\Omega}}(\mathbf{x}, \mathbf{p}_j) = (\boldsymbol{\Omega}(\mathbf{x} - \mathbf{p}_j))^2 \tag{10}$$

with the mapping matrix $\boldsymbol{\Omega} \in \mathbb{R}^{n \times n}$ by SGDL using the gradients

$$\Delta \Omega_{ik} \propto \frac{\partial E(\mathbf{x}, \mathcal{P}, \sigma)}{\partial \Omega_{ik}} \tag{11}$$

of the local error (7) and paying attention to keep $\boldsymbol{\Omega}$ regular [2]. One can easily show that in the expected limit of this adaptation process is $\mathbf{C}^{-1} = \boldsymbol{\Omega}^T \cdot \boldsymbol{\Omega}$.

Remark 2. *If a dimensionality reducing mapping $\boldsymbol{\Omega} \in \mathbb{R}^{m \times n}$ with $m < n$ is used, the SGDL scheme for $\boldsymbol{\Omega}$-training implicitly minimizes the Frobenius-norm $\left\|\boldsymbol{\Omega}^T \cdot \boldsymbol{\Omega} - \mathbf{C}^{-1}\right\|_F$ with $\det(\boldsymbol{\Upsilon}) \neq 0$ for $\boldsymbol{\Upsilon} = \boldsymbol{\Omega} \cdot \boldsymbol{\Omega}^T$ as a regularization constraint [17]. Hence, the row vectors of $\boldsymbol{\Omega}^T$ span the same subspace as the first m eigenvectors of $\mathbf{C}$.*

Using the above covariance estimation, or the corresponding covariance-based pre-processing of the data, the remaining optimization goal of NG/SOM is to distribute the prototypes such that an approximated information-optimal coding is obtained. For NG it is known that for the prototype density ϱ the relation

$$\rho(\mathbf{p}_j) \propto (P(\mathbf{p}_j))^{\alpha}$$

is valid for small but non-vanishing neighborhood range values $0 < \sigma \ll 1$ with $\alpha = \frac{D_{\text{eff}}}{D_{\text{eff}}+2}$ and D_{eff} is the intrinsic dimension of the data and the information optimum coding would be obtained for $\alpha = 1$ [12,16]. Hence, NG is approximately information optimum in case of high-dimensional data.

Remark 3. *For a dimensionality reducing mapping $\boldsymbol{\Omega} \in \mathbb{R}^{m\times n}$ it is recommended to replace the dissimilarity $d_{\boldsymbol{\Omega}}(\mathbf{x}, \mathbf{p}_j)$ from* (10) *by*

$$\delta_{\boldsymbol{\Omega}}(\mathbf{x}, \mathbf{p}_j) = (\boldsymbol{\Omega}\mathbf{x} - \mathbf{p}_j)^2 \tag{12}$$

such that for the prototype set the inclusion $\mathcal{P} \subset \mathbb{R}^m$ is valid. This modification does not influence the behavior and the result of the SGDL for the $\boldsymbol{\Omega}$-adaptation such that Remark 2 remains valid.

3 Integration of Structural Data Knowledge into Adaptive Covariance Learning

Covariance estimation by means of the expectation operator $E_{\mathcal{X}}[\mathbf{x} \cdot \mathbf{x}^T]$ implicitly assumes that further knowledge of the internal data structure is either not available or can be excluded. Otherwise, this assumption is frequently not realistic for application data. For example, for data vectors in gene expression analysis, i.e. genomic profiles, it is a fact that the vector entries x_j representing the gene expression level of a gene, may be related to other expression values x_k by co-dependencies via so-called pathways or related biological processes [6,18], which have been collected and curated in databases like KEGG or Reactome [5,9].

Yet, those structural dependencies are not reflected in the standard covariance analysis. Thus, we propose in the following an approach to integrate such hierarchical knowledge into covariance estimation based on the adaptive scheme provided in Sect. 2.1 for NG and SOM.

For this purpose we assume that external knowledge about possible dependencies between the data vector entries x_j are given as directed graphs, which can be described by layer-wise graphs [17]. We denote these graphs as structure graphs S with layers $L_0, \ldots, L_{h+1}$ where $L_0 = I$ is the input layer and $L_{h+1} = L_{\mathcal{P}}$ is the prototype layer. The number of vertices in the layer L_k is denoted by m_k with $m_0 = n$, $m_h = m$, and $m_{h+1} = C$. Then, the information is transmitted trough the graph along the edges. The transition from one layer $l-1$ to the next layer l can be described by a usually sparse (0-1)-matrix $\mathbf{K}_l \in \{0,1\}^{m_l \times m_{l-1}}$ representing the respective *domain knowledge* regarding the graph structure, where non-zero values code a known edge. Zero values exclude the possibility of a connection. The overall transmission matrix $\mathbf{K}$ for network with $L+1$ layers is obtained as the product $\mathbf{K} = \mathbf{K}_L \cdot \ldots \cdot \mathbf{K}_1 \in \mathbb{R}^{m \times n}$ such that the information flow is given by

$$I = L_0 \xrightarrow{\mathbf{K}_1} L_1 \xrightarrow{\mathbf{K}_2} \ldots \xrightarrow{\mathbf{K}_h} L_h \xrightarrow{\mathbf{D}_{\mathcal{P}}} L_{h+1} = L_{\mathcal{P}}$$

where the matrix $\mathbf{D}_{\mathcal{P}}$ represents the dense connection structure from the last knowledge layer to the prototype layer.

An illustrative example is given in Fig. 1 in the domain of genomic profiles: The input vector entries represent gene expression values related to a set of genes. First, different values x_j in a patients genomic profile are related to the corresponding gene by the matrix $\mathbf{K}_G$. Secondly, all genes that are associated with a biological pathway are forwarded to the next layer by $\mathbf{K}_P$ which is constructed with information available in databases like KEGG or Reactome [5,9]. Simply put, the domain knowledge is coded by connecting edges from the gene layer L_G to the pathway layer L_P in the graph represented by the matrix $\mathbf{K}_P$. Further, pathways may contribute to certain biological processes which, hence, determine the edges from the pathway layer to the biological process layer L_B by means of the transition matrix $\mathbf{K}_B$. From that the information is directly transmitted to the prototype layer using the full matrix $\mathbf{D}_{\mathcal{P}}$, i.e. this information is the essential input to the prototypes in $\mathcal{P}$.

The structural knowledge is not necessarily valid for the given data $\mathcal{X}$, i.e. the experimental conditions, as not all genes, pathways or even biological processes are relevant to it. Instead it provides a set of only possible dependencies given by the expert knowledge. Therefore we assign to each structure matrix $\mathbf{K}_l \in \{0,1\}^{m_l \times m_{l-1}}$ an adjustable matrix $\boldsymbol{\Omega}_l \in \mathbb{R}^{m_l \times m_{l-1}}$ with $[\boldsymbol{\Omega}_l]_{ij} = 0 \Longleftrightarrow [\mathbf{K}_l]_{ij} = 0$ for estimation of valid relations. These matrices serve by $\boldsymbol{\Omega} = \boldsymbol{\Omega}_h \cdot \ldots \cdot \boldsymbol{\Omega}_1$ as trainable parameters for the dissimilarity measures $\delta_{\boldsymbol{\Omega}}(\mathbf{x}, \mathbf{p}_j)$ from (12) or $d_{\boldsymbol{\Omega}}(\mathbf{x}, \mathbf{p}_j)$ from (10) to be used in NG or SOM.

In this way we obtain a biologically-informed prototype based vector quantizer. The prototypes have to be adjusted by SGDL as usual in NG or SOM whereas the matrix entries $[\boldsymbol{\Omega}_l]_{ij}$ of the matrix $\boldsymbol{\Omega}_l$ are adapted using the gradients

$$\frac{\partial \delta_{\boldsymbol{\Omega}}(\mathbf{x}, \mathbf{p})}{\partial [\boldsymbol{\Omega}_l]_{ij}} = \frac{\partial (\boldsymbol{\Omega}_h \cdot \ldots \cdot \boldsymbol{\Omega}_1 \mathbf{x} - \mathbf{p})^2}{\partial [\boldsymbol{\Omega}_l]_{ij}} \quad \text{or} \quad \frac{\partial d_{\boldsymbol{\Omega}}(\mathbf{x}, \mathbf{p})}{\partial [\boldsymbol{\Omega}_l]_{ij}} = \frac{\partial (\boldsymbol{\Omega}_h \cdot \ldots \cdot \boldsymbol{\Omega}_1 (\mathbf{x} - \mathbf{p}))^2}{\partial [\boldsymbol{\Omega}_l]_{ij}}.$$

Interpretation of the knowledge-informed NG/SOM can be easily realized using the *layer-wise dependency matrices* $\boldsymbol{\Lambda}_l$ defined as

$$\boldsymbol{\Lambda}_l = (\boldsymbol{\Omega}_l \cdot \boldsymbol{\Omega}_{l-1} \cdot \ldots \cdot \boldsymbol{\Omega}_1)^T \cdot (\boldsymbol{\Omega}_l \cdot \boldsymbol{\Omega}_{l-1} \cdot \ldots \cdot \boldsymbol{\Omega}_1)$$

such that

$$\boldsymbol{\Lambda}_l = \boldsymbol{\Omega}_l^T \cdot \boldsymbol{\Lambda}_{l+1} \cdot \boldsymbol{\Omega}_l \tag{13}$$

is valid with $\boldsymbol{\Lambda}_{h+1} = \mathbb{E}_n$ being the identity map in the data space $\mathbb{R}^n$ and $\boldsymbol{\Lambda}_h = \boldsymbol{\Lambda} = \boldsymbol{\Omega}^T \boldsymbol{\Omega}$. In the illustrative example, these matrices trace the pathways and biological processes in the biologically-induced structure S to be important

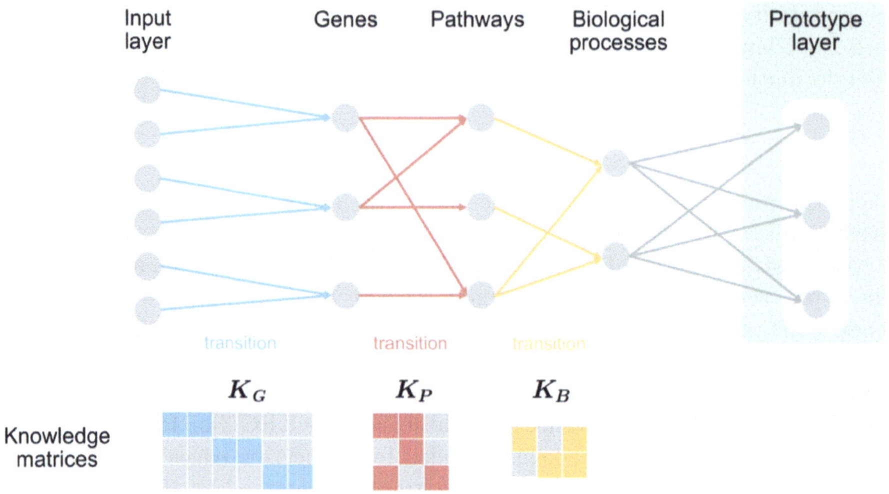

Fig. 1. Generic example of knowledge integration in gene expression analysis. The knowledge matrices $\mathbf{K}_G$, $\mathbf{K}_P$, and $\mathbf{K}_B$ represent the biological knowledge of the layer connections according to the possible dependency structure graph S and, thus, determine the structure of the adjustable matrices $\boldsymbol{\Omega}_1 = \boldsymbol{\Omega}_G$, $\boldsymbol{\Omega}_2 = \boldsymbol{\Omega}_P$, and $\boldsymbol{\Omega}_3 = \boldsymbol{\Omega}_B$ realizing linear maps between the layers. Further, the output of the structural layers serves as the essential input to the prototype layer.

for the given data $\mathcal{X}$. More precisely, the matrix $\boldsymbol{\Lambda}_l$ indicates the layer-wise dependencies between the data features combined up to the layer L_l contributing to an efficient data representation. Further, the (pseudo-) inverse $\mathbf{C}_h$ of $\boldsymbol{\Lambda}_h$ can be interpreted as a knowledge-informed *structured covariance matrix* in dependence on the known biological structure S. Hence, when $\boldsymbol{\Lambda}_h$ is estimated during training the corresponding implicit variance estimation is affected by the structure knowledge.

An important side effect of this knowledge-informed approach is that the matrices $\boldsymbol{\Omega}_l$ are usually very sparse due to the sparse knowledge structure S. This sparseness plays the role of a regularizer and, hence, frequently leads to numerically stable learning.

4 Exemplary Experiments Using Gene Expression Data

4.1 Data Setting and Knowledge Structure

We use the data set together with the domain knowledge which was provided/collected by the authors of [3]. It is a prostate cancer data set and includes genomic profiles of different patients, given by the somatic mutation (changes in the sequence), copy number amplification (increase in a genome fragment) and copy number deletion (missing DNA segment). The domain knowledge, here of biological nature, was derived from the *Reactome* database which groups genes into a hierarchy of increasingly coarse pathways and eventually biological processes. A more detailed description of the data set and the biological background can be found in the paper [3]. The data together with the domain knowledge is publicly available via the links in this paper or directly[1].

We remain with 1,011 patients and the information about mutation, deletion and the copy number variation for 4,739 genes (considering only those for which complete information are available after pre-processing). Further, we identified 1,387 possible pathways and 26 relevant biological processes in the database. To keep the approach as simple as possible in this first demonstration, we did not further decompose the pathways according to their sub-functionalities as it was done in the original investigation [3]. Doing so, we obtain the network structure with input layer $L_0 = I$, three structure layers $L_1 = L_G, L_2 = L_P, L_3 = L_B$ followed by the prototype layer $L_4 = L_{\mathcal{P}}$ as it is illustrated in Fig. 1. Thereby, the possible information flow between the layers we model by the knowledge matrices as introduced in the previous section and here are given as $\mathbf{K}_G \in \{0,1\}^{3\cdot 4739\times 4739}, \mathbf{K}_P \in \{0,1\}^{4739\times 1387}, \mathbf{K}_B \in \{0,1\}^{1387\times 26}$ respectively, derived from the domain knowledge provided by reactome database [5] using the same scheme as it was introduced in [3]. Because we have exactly three values available for each gene in the input vector (mutation, deletion, and copy number variation) the matrix $\mathbf{K}_G$ contains only $3 \cdot 4739$ non-zero entries, which is only 0.02%. For the matrices $\mathbf{K}_p$ and $\mathbf{K}_B$ ratios for non-zero elements are 0.13% and 3.89%, respectively. Hence, all these matrices as well as the corresponding trainable matrices $\boldsymbol{\Omega}_G$, $\boldsymbol{\Omega}_P$, and $\boldsymbol{\Omega}_B$ are sparse such that an inherent regularization during training takes place leading to a stable numerical behavior which frequently is difficult with unrestricted $\boldsymbol{\Omega}$ matrices of this size at least.

4.2 Experiments

For the numerical experiments we apply a neural gas network with 20 prototypes in three different settings (variants):

A NG with metric learning without using the domain knowledge, i.e. unstructured matrix $\boldsymbol{\Omega} \in \mathbb{R}^{3\cdot 4739\times 26}$ is used in the dissimilarity $d_{\boldsymbol{\Omega}}(\mathbf{x}, \mathbf{p}_j)$ from Eq. (10),

B NG with metric learning using the biologically-informed structure matrices, i.e. $\boldsymbol{\Omega} = \boldsymbol{\Omega}_B \cdot \boldsymbol{\Omega}_P \cdot \boldsymbol{\Omega}_G \in \mathbb{R}^{26\times 3\cdot 4739}$ with the sparse knowledge informed matrices $\boldsymbol{\Omega}_B \in \mathbb{R}^{26\times 1387}$, $\boldsymbol{\Omega}_P \in \mathbb{R}^{1387\times 4739}$, and $\boldsymbol{\Omega}_G \in \mathbb{R}^{4739\times 3\cdot 4739}$,

C NG like in **B**, but adding an additional layer to map the data into a two dimensional space (for visualization), i.e. $\boldsymbol{\Omega} = \boldsymbol{\Omega}_V \cdot \boldsymbol{\Omega}_B \cdot \boldsymbol{\Omega}_P \cdot \boldsymbol{\Omega}_G \in \mathbb{R}^{2\times 3\cdot 4739}$ with $\boldsymbol{\Omega}_V \in \mathbb{R}^{2\times 26}$.

Results of the knowledge-informed approach are depicted in Fig. 2 and Fig. 3. We observe that the knowledge-informed structure restriction $\boldsymbol{\Omega} = \boldsymbol{\Omega}_B \cdot \boldsymbol{\Omega}_P \cdot \boldsymbol{\Omega}_G$ of the mapping matrix $\boldsymbol{\Omega}$ leads to a change in the variance structure as shown by way of a principle component analysis of the data with the corresponding

[1] https://drive.google.com/uc?id=17nssbdUylkyQY1ebtxsIw5UzTAd0zxWb.

mappings. Further, applying an additional trained visualization mapping realized by $\boldsymbol{\Omega}_V$, we obtain a more isotropic distribution of the mapped data in $\mathbb{R}^2$ than by PCA. Hence, the knowledge-informed analysis leads to a more appropriate mapping than applying the usual PCA-mapping without structure-adjusted information does. Further, the visualization of the different inverse covariance matrices according to the NG variants **A**, **B**, and **C** as visualized in Fig. 3 emphasize the different behavior of unstructered PCA and structured (inverse) covariance learning within the NG-based representation learning scheme. The respective (inverse) covariance entries of the trained matrices better reflect the biological actualities of the covariates relatedness.

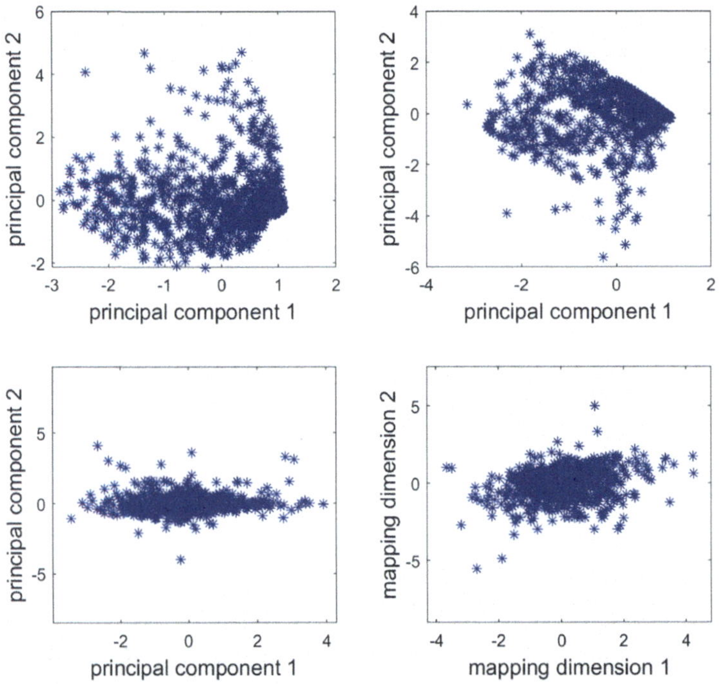

Fig. 2. *upper left*: Visualization of the first two principle components of the original data set applying PCA, *upper right*: the first two principle components of the mapped data using the knowledge-informed structure matrix $\mathbf{K} = \mathbf{K}_B \cdot \mathbf{K}_P \cdot \mathbf{K}_G \in \mathbb{R}^{26\times3\cdot4739}$ applying PCA, *lower left*: the first two principle components of the mapped data using the adjusted knowledge-informed matrix $\boldsymbol{\Omega} = \boldsymbol{\Omega}_B \cdot \boldsymbol{\Omega}_P \cdot \boldsymbol{\Omega}_G \in \mathbb{R}^{26\times3\cdot4739}$ and then applying PCA, *lower right*: mapped data using the mapping matrix learned by neural gas variant **C**, i.e. $\boldsymbol{\Omega} = \boldsymbol{\Omega}_V \cdot \boldsymbol{\Omega}_B \cdot \boldsymbol{\Omega}_P \cdot \boldsymbol{\Omega}_G \cdot \in \mathbb{R}^{2\times3\cdot4739}$

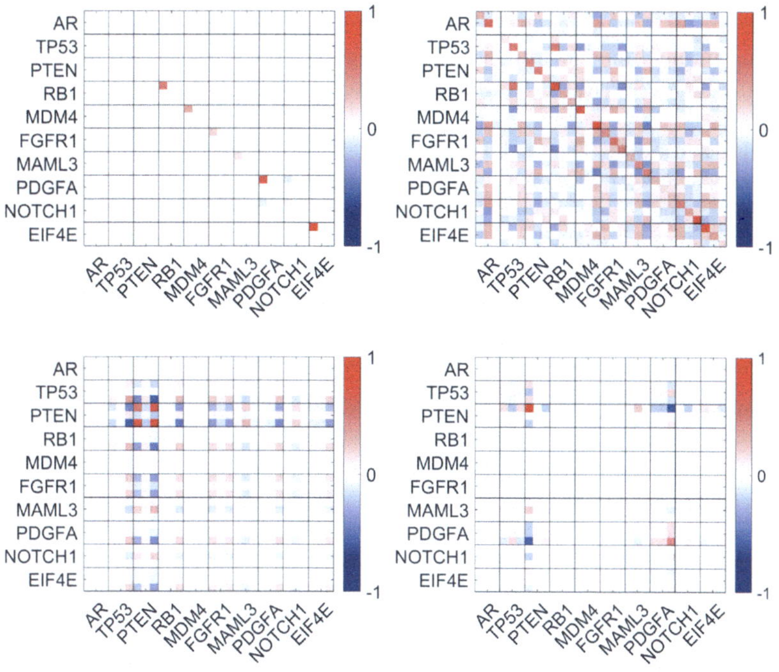

Fig. 3. Visualization of the different variants of inverse covariance matrices reduced to the ten most important genes listed in [3]. For each gene we have three values: mutation, deletion and amplification. *upper left*: reduced inverse covariance matrix of the original data, *upper right*: reduced inverse covariance matrix $\boldsymbol{\Omega}^T\boldsymbol{\Omega}$ according to NG variant **A** without knowledge integration but dimensionality reduced mapping $\boldsymbol{\Omega} \in \mathbb{R}^{26\times 3\cdot 4739}$, *lower left*: reduced inverse covariance matrix $\boldsymbol{\Omega}^T\boldsymbol{\Omega}$ according to NG variant **B** with knowledge integration for the dimensionality reduced mapping $\boldsymbol{\Omega} \in \mathbb{R}^{26\times 3\cdot 4739}$, *lower right*: reduced inverse covariance matrix $\boldsymbol{\Omega}^T\boldsymbol{\Omega}$ according to NG variant **C** with knowledge integration and dimensionality reduced mapping $\boldsymbol{\Omega} \in \mathbb{R}^{2\times 3\cdot 4739}$ to the two-dimensional visualization space.

5 Conclusions

Representation learning by vector quantization requires an appropriate choice of the employed data dissimilarity. Further, covariance analysis by means of principal component analysis gives insights into data structure as well as can be used for dimensionality reduction in addition to representation learning. Yet, if additional knowledge about the data and relations between data attributes is available, a regarding structured covariance analysis is more consistent to the data than an unstructured investigation. In the paper we explained how such an analysis can be realized taking an example from the field of genomic profiles of gene expression analyses.

From the achieved results we can conclude that knowledge-informed data processing in machine learning should be favored in comparison to non-informed methods, which contributes to a better interpretability of the approaches.

Acknowledgements. The authors would like to thank to Stefan Simm (University of Applied Sciences Coburg) and Jan Oldenburg from the Institute of Bioinformatics, University Medicine Greifswald (Germany), for drawing our attention to the promising topic of biology-informed neural networks and providing first insights about this research area to our team. M.K., J. V. and K. S. B. are funded by the German space agency (DLR) - Project AIMS and sub-projects IAI-XPRESS and DAIMLER within the AIMS project.

References

1. Arthur, D., Vassilvitskii, S.: k-means++: the advantages of careful seeding. In: Proceedings of the Eighteenth Annual ACM-SIAM Symposium on Discrete Algorithms, Philadelphia, p. 1. Society for Industrial and Applied Mathematics (2007)
2. Biehl, M., Hammer, B., Schleif, F.M., Schneider, P., Villmann, T.: Stationarity of matrix relevance LVQ. In: Proceedings of the International Joint Conference on Neural Networks 2015 (IJCNN), Los Alamitos, pp. 1–8. IEEE Computer Society Press (2015)
3. Elmarakeby, H., et al.: Biologically informed deep neural network for prostate cancer discovery. Nature **598**, 348–352 (2021). https://doi.org/10.1038/s41586-021-03922-4
4. Erwin, E., Obermayer, K., Schulten, K.: Self-organizing maps: ordering, convergence properties and energy functions. Biol. Cyb. **67**(1), 47–55 (1992)
5. Fabregat, A., et al.: Reactome diagram viewer: data structures and strategies to boost performance. Bioinformatics **34**(7), 1208–1214 (2018). https://doi.org/10.1093/bioinformatics/btx752
6. Hartman, E., et al.: Interpreting biologically informed neural networks for enhanced proteomic biomarker discovery and pathway analysis. Nat. Commun. **14**(5359), 1–13 (2023). https://doi.org/10.1038/s41467-023-41146-4
7. Heskes, T.: Energy functions for self-organizing maps. In: Oja, E., Kaski, S. (eds.) Kohonen Maps, pp. 303–315. Elsevier, Amsterdam (1999)
8. Joliffe, I.: Principal Component Analysis, 2nd edn. Springer, New York (2002). https://doi.org/10.1007/b98835
9. Kanehisa, M., Furumichi, M., Sato, Y., Kawashima, M., Ishiguro-Watanabe, M.: KEGG for taxonomy-based analysis of pathways and genomes. Nucleic Acids Res. **51**, D587–D592 (2023). https://doi.org/10.1093/nar/gkac963
10. Kästner, M., Backhaus, A., Geweniger, T., Haase, S., Seiffert, U., Villmann, T.: Relevance learning in unsupervised vector quantization based on divergences. In: Laaksonen, J., Honkela, T. (eds.) WSOM 2011. LNCS, vol. 6731, pp. 90–100. Springer, Heidelberg (2011). https://doi.org/10.1007/978-3-642-21566-7_9
11. Kohonen, T.: Self-Organizing Maps. Springer Series in Information Sciences, vol. 30. Springer, Heidelberg (1995). (Second Extended Edition 1997)
12. Martinetz, T.M., Berkovich, S.G., Schulten, K.J.: 'Neural-gas' network for vector quantization and its application to time-series prediction. IEEE Trans. Neural Networks **4**(4), 558–569 (1993)

13. Nebel, D., Kaden, M., Villmann, A., Villmann, T.: Types of (dis−)similarities and adaptive mixtures thereof for improved classification learning. Neurocomputing **268**, 42–54 (2017). https://doi.org/10.1016/j.neucom.2016.12.091
14. Schneider, P., Hammer, B., Biehl, M.: Distance learning in discriminative vector quantization. Neural Comput. **21**, 2942–2969 (2009)
15. Tversky, A.: Features of similarity. Psychol. Rev. **84**(4), 324–352 (1977)
16. Villmann, T., Claussen, J.C.: Magnification control in self-organizing maps and neural gas. Neural Comput. **18**(2), 446–469 (2006)
17. Voigt, J., Saralajew, S., Kaden, M., Reuss, L., Villmann, T.: Biologically-informed shallow classification learning integrating pathway knowledge. In: Proceedings 17th International Joint Conference on Biomedical Engineering Systems and Technologies (BIOSTEC 2024) – Bioinformatics Workshop. p. accepted. SCITEPRESS - Science and Technology Publications, LDA (2024)
18. Wysocka, M., Wysocki, O., Zufferey, M., Landers, D., Freitas, A.: A systematic review of biologically-informed deep learning models for cancer: fundamental trends for encoding and interpreting oncology data. BMC Bioinform. **24**(198), 1–31 (2023). https://doi.org/10.1186/s12859-023-05262-8
19. Zador, P.L.: Asymptotic quantization error of continuous signals and the quantization dimension. IEEE Trans. Inf. Theory **IT-28**, 149–159 (1982)

Exploring Data Distributions in Machine Learning Models with SOMs

Caroline König(✉) and Alfredo Vellido

Intelligent Data Science and Artificial Intelligence (IDEAI-UPC) Research Centre,
Universitat Politècnica de Catalunya (UPC Barcelona Tech),
Jordi Girona 1-3, 08034 Barcelona, Spain
{ckonig,avellido}@cs.upc.edu

Abstract. Data quality control is fundamental in data-driven analysis with machine learning (ML) models. In the domain of drug research, there is an increasing interest in the prediction of relevant biocompounds physicochemical properties with ML. In order to build predictive models of good quality, it is important to adequately select representative datasets. In this work, we combine ML prediction and Self-Organizing Maps-based exploration to build an interpretable machine learning model and to characterize those data that are most difficult to predict in the validation stage.

Keywords: Self-Organizing Maps · ADME properties · Random Forest · Interpretable Machine Learning · Data Quality · Bioinformatics

1 Introduction

The concept of data quality control is key in computational biomedicine in general and in bioinformatics in particular. Data quality is in turn fundamental for the successful application of machine learning (ML) models at large [6], at a time in which such methods have become commonplace in the bioinformatics domain, for problems such as the prediction of physicochemical properties of biomolecules. In the area of drug design, the assessment of the so-called ADME properties (absorption, distribution, metabolism, and excretion) is of particular relevance, as such properties determine the effectiveness of the drug on-site. Over the last decades, there have been advances in the development of ML-based predictive models for quantitative structure-property/activity relationships (QSPRs/QSARs) in general [7,17] and for ADME property prediction in particular [1,4,12]. This progress is partially related to the release of public available molecular databases, such as [3], which provides a curated dataset for the prediction of six internal in vitro ADME endpoints. Nevertheless, in the field of molecular property prediction, the representation of features is still an open challenge. In [3], the development of ML models was investigated assuming differentiated features, with either Functional Connectivity Fingerprints or molecular descriptors, or a hybrid combination of both molecular representations.

T. Villmann et al. (Eds.): WSOM+2024, LNNS 1087, pp. 86–95, 2024.
https://doi.org/10.1007/978-3-031-67159-3_10

In this study, we aim to provide an exploratory analysis of the data used for training and validation of an ML prediction model for this problem, so as to understand the distribution of data used for training and to discover patterns in those observations that are difficult to predict through ML. For this, we investigate which structural features are most relevant for the activity prediction and then we use exploratory analysis with Self Organizing Maps (SOM), in a mixed visualization and clustering approach, to understand the distribution of data in that reduced dataset. The remainder of this brief paper is structured as follows. Section 2 describes the data under study, Sect. 3 describes the ML approaches for modelling, feature relevance analysis and exploratory analysis. Finally, Sects. 4 and 5 present and discuss the results of the experiments.

2 Materials

The dataset under study, released by [3], is used for the prediction of the ADME properties of a compound, which are critical to assess the drug effectiveness at the site of action. Although the original study provides data for the prediction of six ADME endpoints, our investigation is limited, for the sake of brevity, to the data for the prediction of human liver microsomal (HLM) stability. The dataset provides a training and validation set comprising respectively 2469 and 648 observations and 316 molecular descriptors calculated with the RDKIt library[1].

Each molecular compound is specified by the Simplified Molecular-Input Line Entry (SMILE) notation, which is based on ASCII strings and describes the molecule's structure. Additionally, the 3D model of the compound is also provided, specifying the 3D position of each atom. Nevertheless, the physicochemical properties of a compound are not directly retrievable from the SMILE or the molecular graph representation. Hence, the calculation of molecular descriptors is needed for quantifying the physicochemical properties.

3 Methods

A predictive model for the HLM activity dataset is built using Random Forest (RF) regression, a decision tree-ensemble model [2] that has been reported to perform well in previous studies [3]. Its performance in the regression task is measured using the Root Mean Squared Error (RMSE). The ensemble model is trained with several RF regressors, using 3-fold cross-validation on the training dataset, and its performance assessed on a validation set.

Furthermore, a feature relevance analysis is carried out to estimate the most relevant features for the prediction of activity. For this, feature permutation is implemented in the *caret* package's RF implementation in R[10], which includes both the feature permutation method and the Gini index approach. In the following, an RF model is trained repeatedly based on a subset of relevant features to verify if regression performance remains stable with a reduced set of features.

[1] https://www.rdkit.org/.

Once the best-performing RF model and the related set of relevant features is found, an unsupervised exploratory analysis is carried out to explore the characteristics of both the training and validation sets. Regarding the training dataset, it is of interest to discover the sample heterogeneity of distribution regarding the most important molecular properties for the prediction model. Regarding the validation dataset, it is of interest to discover the differences across observations according to their prediction error. While the training set comprises 2468 samples, the validation set comprises only 618 samples, which are both explored with SOM [8], an unsupervised artificial neural network that aims to construct lower-dimensional (amenable to visualization) data representations in the form of an interpretable mapping of the higher-dimensional entry data. In this study, the SOM implementation of the *SOMbrero* package [14] was used.

4 Results

4.1 Regression Models and Feature Relevance Analysis

Figure 1 describes the performance of the RF models built on the different number of features in the prediction of the HLM activity. The baseline RF model was built on the whole 316 molecular descriptor properties and yielded an RMSE of 0.49 on the training dataset (using 3-fold cross-validation) and an RMSE of 0.4 on the additional validation set provided by the datasets authors. Based on this baseline model, feature relevance was assessed using the feature permutation implemented with RF. Figure 1b shows the rank of the most relevant features for the prediction of the HLM activity. They are the partition coefficient (logP value) calculated by the Wildman-Crippen descriptor [16] (V1148), the Hall-Kier kappa2 molecular shape descriptor [11] (V1045), three atomic polarizability (V1181,V1180,1270) and one relative IState (V1292) descriptor from the 2D autocorrelation descriptor, the topological polar surface area (TPSA) descriptor (V1026), the Ch1v connectivity and shape index [11] (V1050), the molecular quantum number descriptor [13] (V1124) and the Inertial Shape Factor descriptor (V1059) [15].

The iterative construction of RF ensemble models on a reduced set of relevant features yields models with similar performance according to the RMSE value on the training and validation set as described in Fig. 1a.

For each RF model built on a different feature set, a SOM model was constructed and its quality was evaluated using the topographic error (which describes the percentage of observations for which the best-matching unit -BMU- is not a neighbor of the second-BMU on the map; that is, it measures topology preservation, so that 0 indicates maximum projection quality of projection, while 1 means minimum quality). The number of features for building the SOMs was augmented with an additional feature: the HLM activity value predicted by the RF models. The topographic error values are 0.217 for 317 descriptors, 0.089 for 51, 0.090 for 21, 0.128 for 11 and 0.027 for 6. Hence, the SOM built on the most parsimonious dataset (6 features) is used in the following for the exploratory analysis of the training and validation dataset.

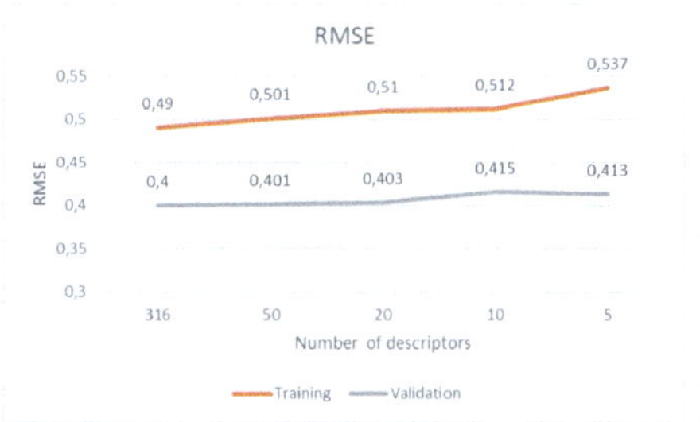

(a) RMSE for training and validation sets as a function of the number of features.

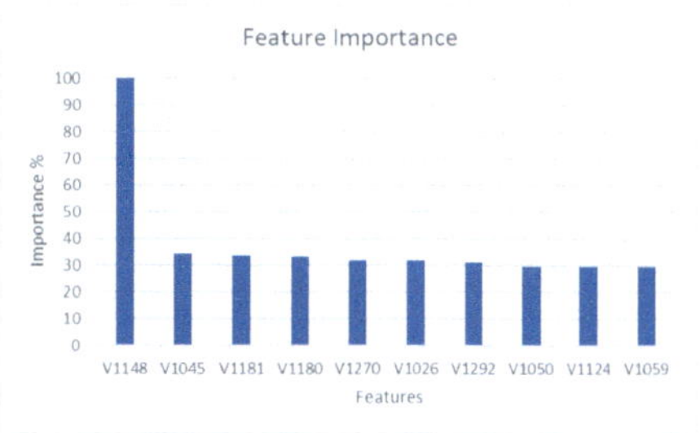

(b) Feature relevance for the prediction of the HLM activity.

Fig. 1. (a) RMSE for training and validation sets as a function of the number of features. (b) Feature relevance for the prediction of the HLM activity.

4.2 Exploratory Analysis of the Training Dataset

The SOM constructed from the training dataset comprises 2469 samples and 6 features, which are projected on a 4×4 grid, yielding a distribution into 16 clusters. This grid size is considered as a reasonable choice for the problem at hand, as we are interested in sizeable cluster solutions. Using hierarchical clustering, these 16 clusters can be further grouped into 5 super-clusters as shown in Fig. 2a. The cluster sizes range between 105 and 205 observations. Figure 2b displays the feature values profiles for the prototypes of all 16 clusters. This representation also allows us to appreciate the clear similarities between prototypes belonging to the same super-cluster, as depicted in the SOM grid of Fig. 2c. The specific distribution of the activity property is shown in Figs. 2d and 2e, overlaid on the SOM grid and using, in turn, color-coded and 3D perspective representations. This activity feature is seen to evolve smoothly over the grid of SOM units, mapping low values onto the lower left corner and high values onto the upper right corner.

An analysis of the distribution of all molecular descriptor features across the prototypes of the SOM grid again reveals a smoothly changing distribution of values, as represented in Fig. 3.

4.3 Exploratory Analysis of the Validation Set

Another SOM model was created for the validation dataset comprising the five most relevant molecular descriptor properties, plus the activity property and the RMSE obtained by the RF model. The information about the RMSE is added as the goal of the exploratory analysis is the discovery of the characteristics of compounds that are more difficult for the RF model to predict (that is, those with higher RMSE).

Figure 4c shows the clustering of this augmented validation dataset of 648 observation onto a grid of 16 clusters, with resulting cluster sizes ranging between 27 and 58 observations. Figure 4a further reveals the grouping into five more

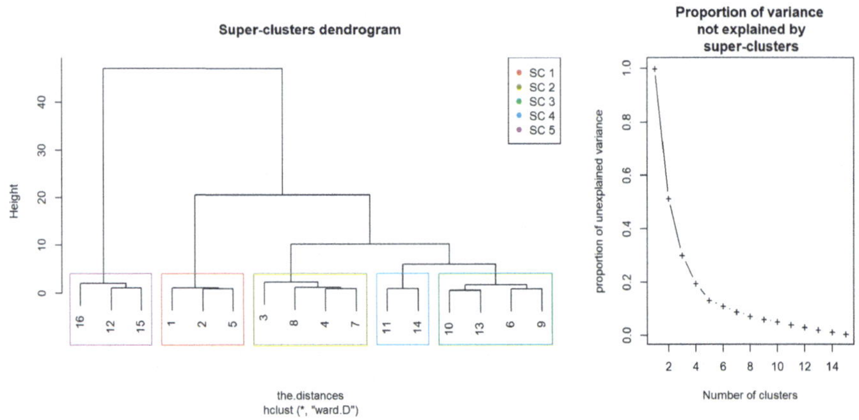

(a) Super-clusters derived from the 4x4 grid SOM with hierarchical clustering.

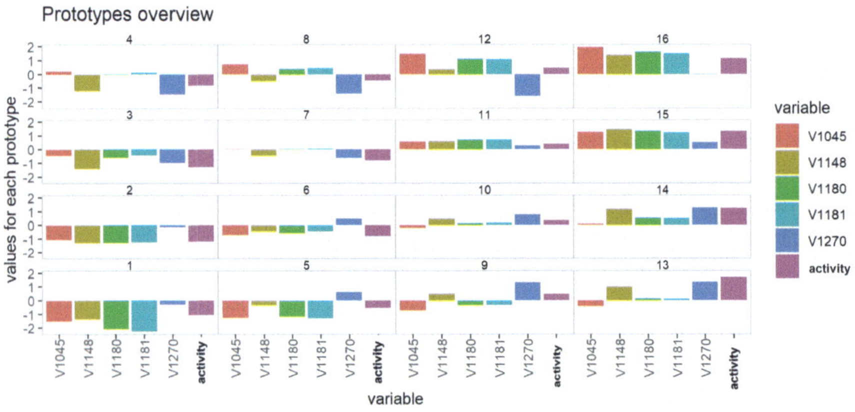

(b) Graphical representation of the features of each prototype.

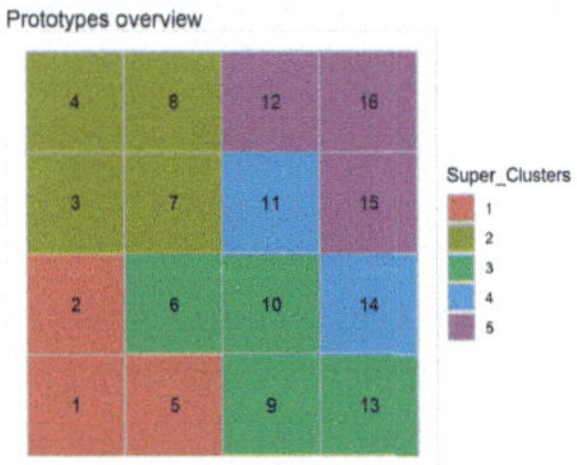

(c) Clusters

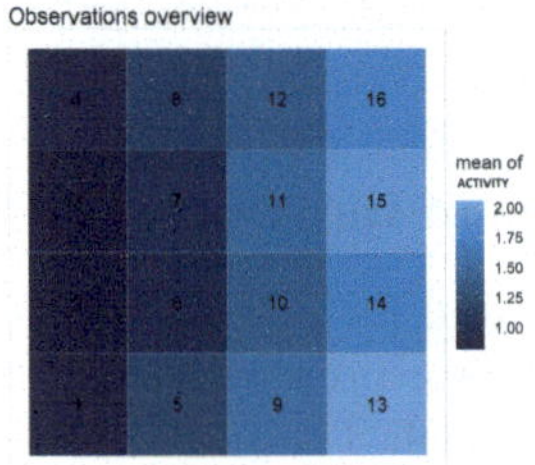

(d) Grid distribution

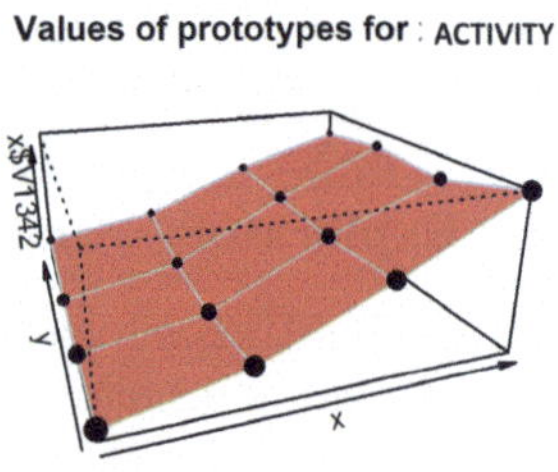

(e) 3D representation

Fig. 2. SOM of the training data.

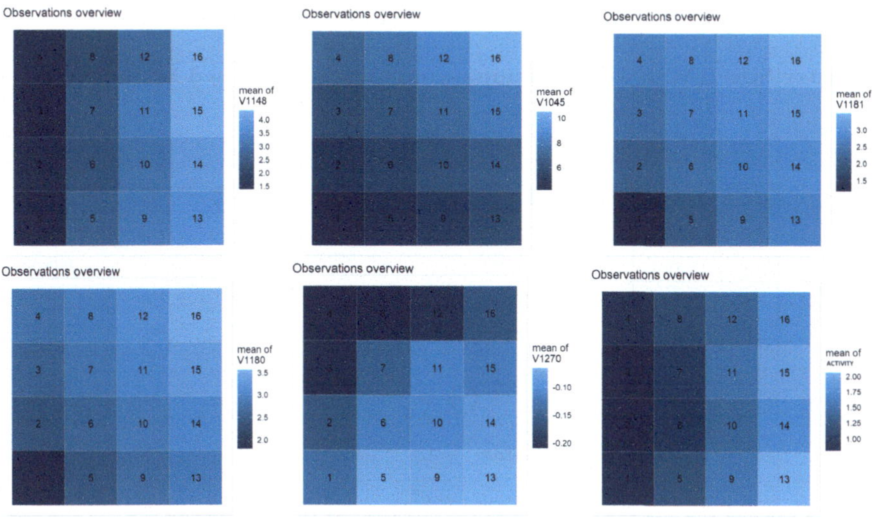

Fig. 3. Feature distribution over the grid training set.

generic groups through hierarchical clustering. Figure 4b shows the feature value profiles for the prototype of each cluster. Furthermore, the distribution of the RSME over the SOM prototype grid layout is shown, in turn, in colour coded and 3D perspective representations, in Figs. 4d and 4e. The distribution of the RMSE over the grid is less smooth and more heterogeneous than the other features (Fig. 5). There are 5 clusters describing groups of observations with a higher RMSE value compared to the mean RMSE of the validation dataset of 0.44. Table 1 describes, for each cluster, the number of observations (size), the value of RMSE and of the HLM activity of the prototype, as well as three examples of compounds of the cluster. The representativeness of the selected compounds is based on their similarity to the properties of the cluster prototype.

5 Discussion

The projection of the observed data produced by SOM yielded good results for both the training and validation sets. In both cases, the samples were distributed quite homogeneously between groups, even without an exhaustive parameter fine-tuning, suggesting that both datasets do not include atypical data or outliers. Moreover, the dataset becomes easily interpretable as the SOM representation provides a cluster description based on the feature profile of the obtained prototypes, which we have graphically described here with barplots (See Fig. 2b and Fig. 4b).

In the training set, the distribution of features evolves very smoothly and almost linearly over the grid, whereas the analysis of the validation dataset including the RMSE features showed some nonlinear distribution over the grid.

Table 1. Cluster analysis with regard to high RMSE values

Cluster	Size	RMSE	HLM	Sample compounds described by vendor Id		
1	58	0.484	1.85	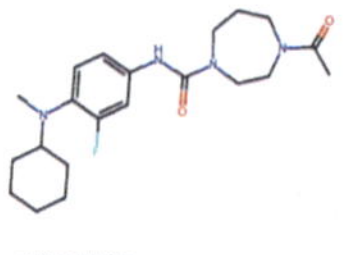 32015702	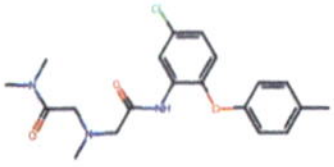 44014989	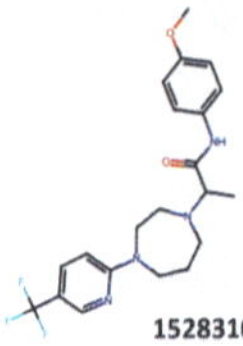 15283106
2	38	0.51	1.59	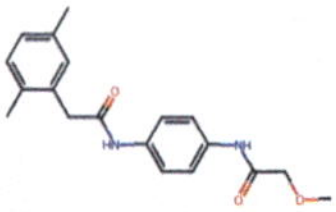 232900428	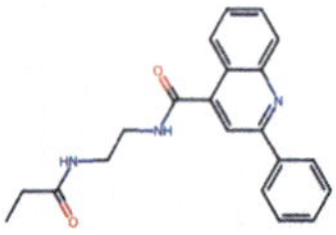 5965122	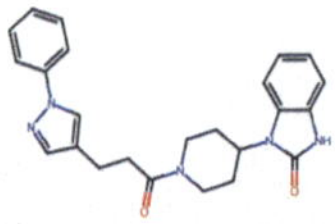 15272357
5	36	0.521	1.90	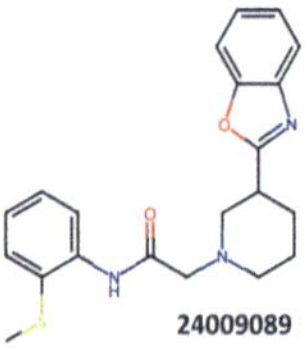 24009089	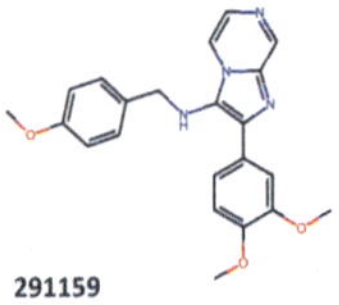 291159	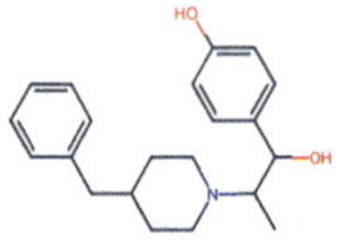 1933590
6	36	0.518	1.47	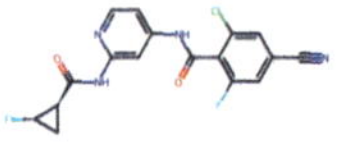 313524054	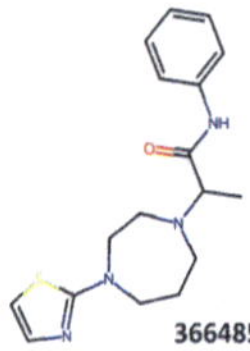 36648529	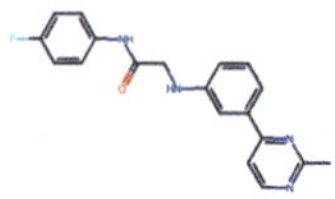 24031550
10	36	0.482	1.36	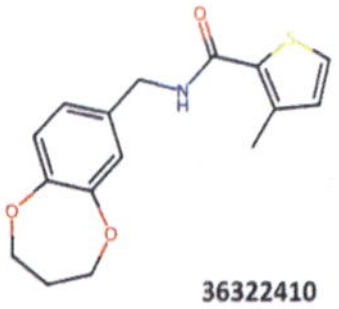 36322410	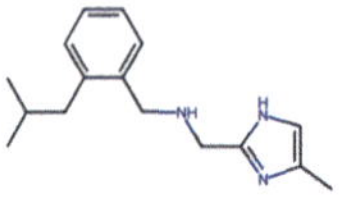 276241705	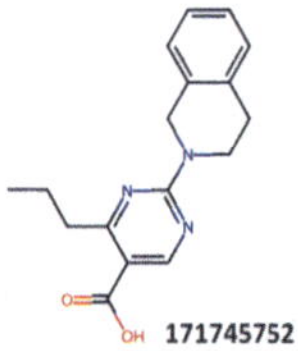 171745752

This insight is interesting, as it highlights that the difficulty of prediction is not directly related to a given type of compound: high RMSE values were found in different and not adjacent clusters (See Fig. 4e).

The analysis of the characteristics of the clusters with a high RMSE provides further insights about the characteristics of its compounds, which can be useful to improve the dataset for training. It is important to recall that known sampling methods to improve data quality are not straightforwardly applicable to molecular data, as the feature set is derived in a non-bidirectional manner by calculating molecular descriptors from the compound. For this reason, more complex approaches for compound selection to improve diversity for molecular

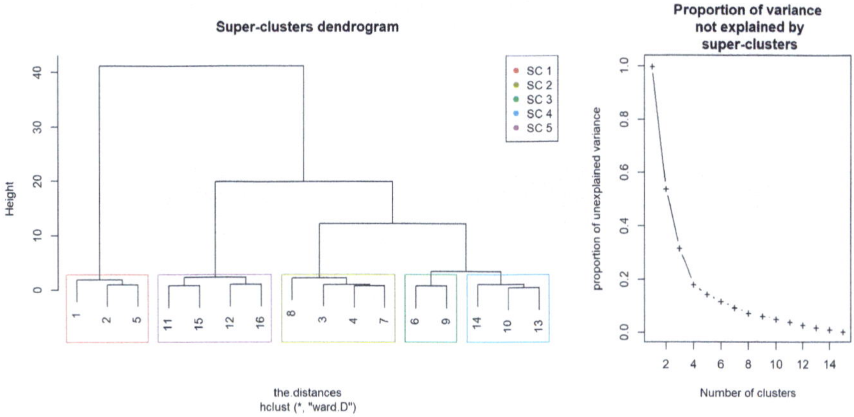

(a) Supercluster derived with hierarchical clustering from the 4x4 grid SOM.

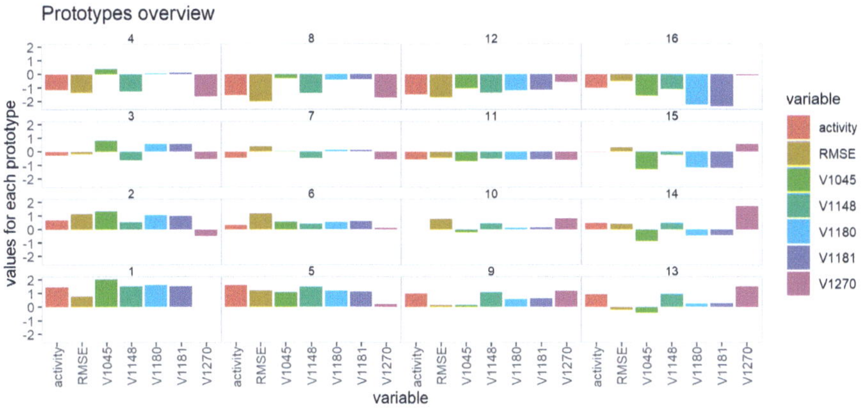

(b) Graphical representation of the features of each prototype.

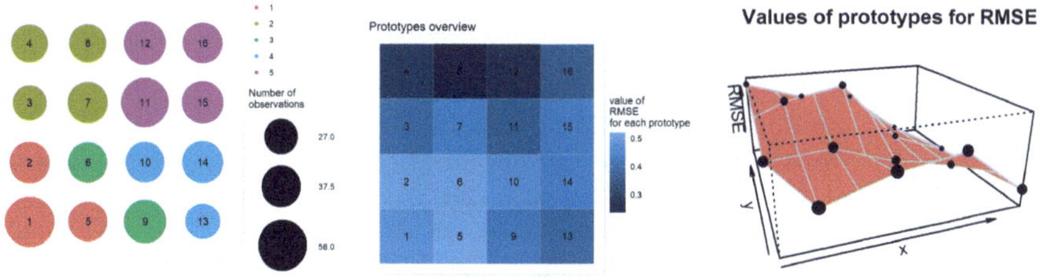

(c) Clusters (d) RMSE distribution. (e) 3D projection RMSE

Fig. 4. SOM of the validation data.

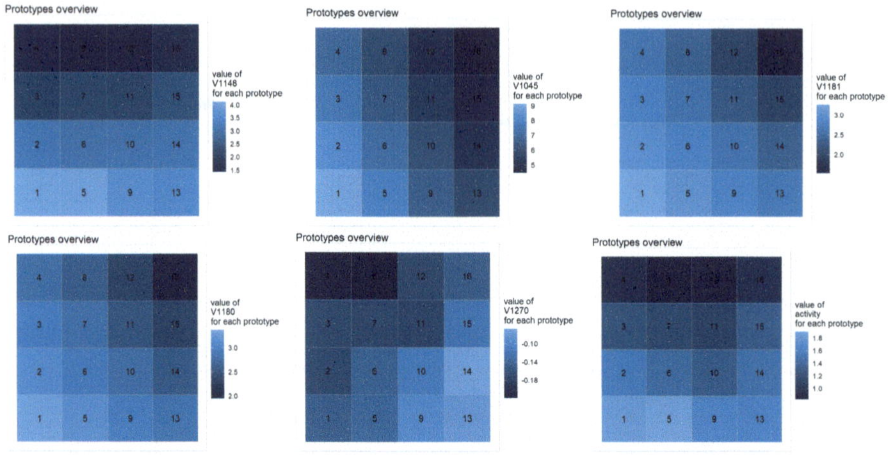

Fig. 5. Feature distribution over the grid validation set.

data are needed [9]. Furthermore, the information provided by this study, the insights about the compounds in the clusters that were more difficult to predict by the model (Table 1) are of interest for the experts because this information enables them to improve iteratively the training dataset by understanding the characteristics of the compounds which resulted more difficult to predict by the model.

In future research, we plan to extend the analysis of the ADME property prediction to all six ADME properties provided by the original study [3] and to use hybrid SOMs, such as Learning-regulated Context Relevant Maps[5] for example, which pair an interpretable SOM front-end directly with a classifier or regression back-end.

Acknowledgements. This research was partially funded by Spanish AEI PID2022-143299OB-I00 grant.

References

1. Balani, S.K., Miwa, G.T., Gan, L.S., Wu, J.T., Lee, F.W.: Strategy of utilizing in vitro and in vivo ADME tools for lead optimization and drug candidate selection. Curr. Top. Med. Chem. **5**(11), 1033–1038 (2005)
2. Breiman, L.: Random forests. Mach. Learn. **45**, 5–32 (2001)
3. Fang, C., et al.: Prospective validation of machine learning algorithms for absorption, distribution, metabolism, and excretion prediction: an industrial perspective. J. Chem. Inf. Model. (2023)
4. Feinberg, E.N., Joshi, E., Pande, V.S., Cheng, A.C.: Improvement in ADMET prediction with multitask deep featurization. J. Med. Chem. **63**(16), 8835–8848 (2020)

5. Hartono, P., Hollensen, P., Trappenberg, T.: Learning-regulated context relevant topographical map. IEEE Trans. Neural Netw. Learn. Syst. **26**(10), 2323–2335 (2014)
6. Jain, A., et al.: Overview and importance of data quality for machine learning tasks. In: Proceedings of the 26th ACM SIGKDD International Conference on Knowledge Discovery & Data Mining, pp. 3561–3562 (2020)
7. Keyvanpour, M.R., Shirzad, M.B.: An analysis of QSAR research based on machine learning concepts. Curr. Drug Discov. Technol. **18**(1), 17–30 (2021)
8. Kohonen, T.: The self-organizing map. Proc. IEEE **78**(9), 1464–1480 (1990)
9. Koutsoukas, A., et al.: How diverse are diversity assessment methods? A comparative analysis and benchmarking of molecular descriptor space. J. Chem. Inf. Model. **54**(1), 230–242 (2014)
10. Kuhn, M.: Building predictive models in R using the caret package. J. Stat. Softw. **28**, 1–26 (2008)
11. Lipkowitz, K.B., Boyd, D.B.: Reviews in Computational Chemistry 6. Wiley Online Library (1995)
12. Ma, J., Sheridan, R.P., Liaw, A., Dahl, G.E., Svetnik, V.: Deep neural nets as a method for quantitative structure-activity relationships. J. Chem. Inf. Model. **55**(2), 263–274 (2015)
13. Nguyen, K.T., Blum, L.C., Van Deursen, R., Reymond, J.L.: Classification of organic molecules by molecular quantum numbers. ChemMedChem Chem. Enabling Drug Discov. **4**(11), 1803–1805 (2009)
14. Olteanu, M., Villa-Vialaneix, N.: On-line relational and multiple relational SOM. Neurocomputing **147**, 15–30 (2015)
15. Todeschini, R., Consonni, V.: Descriptors from Molecular Geometry, chap. VIII.2, pp. 1004–1033. Wiley (2003)
16. Wildman, S.A., Crippen, G.M.: Prediction of physicochemical parameters by atomic contributions. J. Chem. Inf. Comput. Sci. **39**(5), 868–873 (1999)
17. Wu, Z., et al.: Do we need different machine learning algorithms for QSAR modeling? A comprehensive assessment of 16 machine learning algorithms on 14 QSAR data sets. Briefings Bioinform. **22**(4), bbaa321 (2020)

Interpretable Machine Learning in Endocrinology: A Diagnostic Tool in Primary Aldosteronism

Michael Biehl[1,2](✉), David Pavlov[1], Alice J. Sitch[3,4], Alessandro Prete[3,5,6], and Wiebke Arlt[7,8]

[1] Bernoulli Institute for Mathematics, Computer Science and Artificial Intelligence, University of Groningen, Groningen, The Netherlands
m.biehl@rug.nl
[2] Centre for Systems Modelling and Quantitative Biomedicine, University of Birmingham, Birmingham, UK
[3] NIHR Birmingham Biomedical Research Centre, University of Birmingham and University Hospitals Birmingham NHS Foundation Trust, Birmingham, UK
{a.j.sitch,a.prete}@bham.ac.uk
[4] Institute of Applied Health Research, University of Birmingham, Birmingham, UK
[5] Institute of Metabolism and Systems Research, College of Medical and Dental Sciences, University of Birmingham, Birmingham, UK
[6] Centre for Endocrinology, Diabetes and Metabolism, Birmingham Health Partners, Birmingham, UK
[7] Medical Research Council Laboratory of Medical Sciences, London, UK
w.arlt@lms.mrc.ac.uk
[8] Institute of Clinical Sciences, Faculty of Medicine, Imperial College London, London, UK

Abstract. The use of urinary steroid metabolomics (USM) in combination with machine learning in endocrinology is briefly introduced. We demonstrate the usefulness of the approach for the detection and differential diagnosis of Primary Aldosteronism (PA), which has been addressed in a recent retrospective study. Here, we mainly present results for the application of the prototype based Generalized Matrix Relevance Learning Vector Quantization for the classification of steroid metabolomics profiles. The method allows for the successful diagnosis of PA and provides insights into the importance of the available markers. Moreover, it facilitates the non-invasive identification of a subtype of PA which is associated with adrenal adenoma harboring a particular mutation in the tumor tissue.

Keywords: Interpretable Machine Learning · Steroid Metabolomics · Primary Aldosteronism

1 Introduction, Background and Motivation

The adrenal glands, located on top of the kidneys, play a major role in the human endocrine system. They produce steroid hormons and metabolites which

T. Villmann et al. (Eds.): WSOM+2024, LNNS 1087, pp. 96–105, 2024.
https://doi.org/10.1007/978-3-031-67159-3_11

regulate, for instance, the immune system, metabolism or sexual development. Consequently, the presence of tumors in the adrenal glands or other disorders can affect a wide range of body functions.

In recent years, Urinary Steroid Metabolomics (USM) has been put forward and established as a useful framework in this context, see e.g. [2,21]. The non-invasive analysis of steroid metabolite excretion in the urine, in combination with interpretable, prototype-based machine learning techniques [3], has facilitated the development of novel diagnostic tools and provided valuable insights into disease mechanisms. Previous applications of USM addressed, for instance, the detection of malignancy in adrenocortical tumors [2,4,22], tumor recurrence after treatment [10], or the differential diagnosis of liver disorders [15].

In this contribution, we consider the use of USM as a diagnostic tool in Primary Aldosteronism (PA). This condition is characterized by an overproduction of aldosterone which plays an important role in the regulation of the blood pressure. PA is known to be the most common form of secondary hypertension and is associated with an increased risk for cardiovascular complications and other comorbidities, such as type 2 diabetes and osteoporosis.

Several subtypes of PA have been identified which reflect the origin of the disorder: The main cause of so-called unilateral PA (UPA) is the presence of an aldosterone producing adenoma (APA) in one of the glands. Subtypes of APA correspond to a number of different driver mutations that can be found in the tumor tissue. In contrast, bilateral PA (BPA) are most frequently due to hyperplasia of both glands, i.e. the excessive growth of otherwise normal adrenocortical tissue.

Currently, the precise diagnosis of PA and its subtypes is non-trivial, time-consuming and complex. As a consequence, it can be assumed that a significant number of cases remain undiagnosed, resulting in, for instance, a poor choice of treatment. In [18] we demonstrated, in collaboration with an extended list of co–authors, that USM bears the promise to provide a diagnosis tool that can overcome part of the diagnostic challenge. To this end, steroid excretion data in a cohort of Healthy Controls (HC) and patients with various forms of PA were analysed by use of interpretable machine learning techniques.

The medical background and implications of our work are discussed in greater detail in [18]. Here we focus on the machine learning aspects of the study. In the following section we briefly summarize characteristics of the patient cohort and the data acquisition. Thereafter, we provide the basic workflow and details of the machine learning analyses in Sect. 3. Key results are presented and discussed in Sect. 4, before we conclude and suggest follow-up projects in Sect. 5.

2 The Data

In this study, 65 healthy controls (HC) and 158 patients diagnosed with Primary Aldosteronism (PA) were included. For all subjects, 24 hour urine samples were analysed by gas chromatography-mass spectroscopy (GCMS), as described in [2] in greater detail. Excretion values for a panel of 34 steroid metabolites were

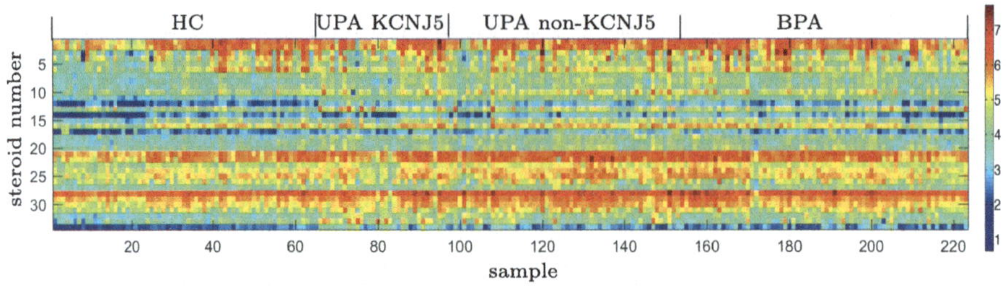

Fig. 1. Color-coded representation of the 34 log-transformed excretion values in 223 samples, grouped according to class membership, see Sect. 2 for details.

obtained, including mineralocoricoids, glucocorticoids, hybrid steroids, androgens, and the corresponding precursors, see e.g. [2,21] for details and further references.

In Sect. 4 we will provide the names of specific metabolites which appear to play an essential role in the different conditions. For further details and a full list of the metabolites considered here, we refer the reader to [18] and to [21] for nomenclature.

The patient cohort comprised 70 patients with bilateral PA (BPA) and 88 cases of unilateral PA (UPA). For the latter group, targeted gene sequencing was performed on extracted tumor tissue, which revealed that 31 samples harbored the particular somatic mutation KCNJ5. For the remaining 57 UPA samples, other somatic mutations were found or no mutation was identified in the tumor tissue. Again, we refer the reader to [18] for further details and references.

Due to the skewed distribution of observations, displaying many small and few relatively large values, we applied a logarithmic transformation of the excretion values. This has proven beneficial in previous steroide metabolomics studies, see e.g. [2,15].

Figure 1 displays the log-transformed excretion values of 34 steroid metabolites in color code. Samples are grouped according to the relevant subgroups, i.e. HC, UPA (with mutation KCNJ5), UPA (non-KCNJ5), and BPA.

3 Machine Learning Analysis

As outlined in the previous section, $N = 34$ log-transformed steroid metabolite excretion values were obtained per sample. The resulting concatenated values were interpreted as N–dimensional feature vectors $\mathbf{x}^\mu$. The index μ specifies the sample (subject) and labels $y^\mu \in \{1, 2, \ldots C\}$ indicate the class memberships. Here, we restrict ourselves to binary diagnosis problems with $C = 2$. Specifically, we considered the discrimination of *HC vs. PA*, *UPA vs. BPA* and *KCNJ5 vs. non-KCNJ5* samples. For each of these classification problems, we constructed a data set of the form

$$\mathbb{D} = \{\mathbf{x}^\mu, y^\mu\}_{\mu=1}^{P} \text{ with feature vectors } \mathbf{x}^\mu \in \mathbb{R}^N \text{ and labels } y^\mu \in \{1, 2\},$$

where P denotes the corresponding number of available samples.

The application of an interpretable, prototype-based system for the classification was in the center of our interest [3,6]. In the following subsection, we briefly summarize essentials of the applied training prescription. For comparison, we also applied and analysed a Random Forest (RF) classifier to the same classification problems [9].

3.1 Generalized Matrix Relevance Learning Vector Quantization

Learning Vector Quantization was introduced by Teuvo Kohonen as a framework for the prototype-based supervised learning of classification schemes [3,6,13, 16]. The central idea is to identify prototype vectors $\mathbf{w}^j$ in feature space which represent the classes:

$$\left\{ \mathbf{w}^j, c^j \right\}_{j=1}^{M} \text{ where } \mathbf{w}^j \in \mathbb{R}^N \text{ and } c^j \in \{1, 2, \ldots C\}. \tag{1}$$

In the simplest settings $M = C$, i.e. each class is represented by exactly one prototype. Together with a non-negative distance measure $d(\mathbf{w}^j, \mathbf{x})$, the simple, so-called Nearest Prototype Classifier (NPC) assigns an arbitrary feature vector $\mathbf{x} \in \mathbb{R}^N$ to the class

$$y(\mathbf{x}) = c^* \text{ where } c^* \text{ is the label of } \mathbf{w}^*(\mathbf{x}) = \text{argmin}_{\mathbf{w}^{\mathrm{j}}} \left\{ \mathrm{d}(\mathbf{x}, \mathbf{w}^{\mathrm{j}}) \right\}_{\mathrm{j}=1}^{\mathrm{M}}. \tag{2}$$

The term *winner* is commonly used for the closest prototype $\mathbf{w}^*(\mathbf{x})$ for a given feature vector [3]. Modified classification schemes can be obtained by considering *soft* or probabilistic voting schemes replacing Eq. (2).

A variety of LVQ training algorithms have been suggested in the literature, see e.g. [16] for a review. A popular cost function based approach Generalized LVQ (GLVQ) [19] is a popular approach which is guided by the minimization of

$$E = \sum_{\mu=1}^{Q} \Phi(e^\mu) \quad \text{with } e^\mu = \frac{d(\mathbf{w}_\mu^J, \mathbf{x}^\mu) - d(\mathbf{w}_\mu^K, \mathbf{x}^\mu)}{d(\mathbf{w}_\mu^J, \mathbf{x}^\mu) + d(\mathbf{w}_\mu^K, \mathbf{x}^\mu)}. \tag{3}$$

Here, the sum is over all Q example data in the actual training set. For a given $\mathbf{x}^\mu$, $\mathbf{w}_\mu^J$ denotes the closest of all prototypes which carry the same label as the example, while $\mathbf{w}^K$ is the closest prototype with a label different from y^μ. The simple identity function $\Phi(z) = z$ has been used throughout this work. Note that the contribution of a single example obeys $-1 \leq e^\mu \leq 1$ and that negative values $e^\mu < 0$ correspond to a data point which is correctly classified by the NPC.

Extensions of GLVQ including so-called relevance learning schemes have been introduced in e.g. [12,20]. In particular, Generalized Matrix Relevance LVQ (GMLVQ) [20] has been applied successfully in a variety of practical problems, see e.g. [1,2,7,8,23], GMLVQ employs a distance measure of the form

$$d(\mathbf{w}^j, \mathbf{x}) = \left(\mathbf{x} - \mathbf{w}^j\right)^\top \Lambda \left(\mathbf{x} - \mathbf{w}^j\right) \tag{4}$$

with the so-called relevance matrix $\Lambda \in \mathbb{R}^{N\times N}$ of adaptive parameters. The reparameterization $\Lambda = \Omega^\top \Omega$ guarantees non-negativity of the measure and implies the symmetry $\Lambda_{ij} = \Lambda_{ji}$. In the following we assume that also $\Omega \in \mathbb{R}^{N\times N}$ and for the sake of numerical stability impose a normalization of the form [20]

$$\sum_{j=1}^{N} \Lambda_{jj} = \sum_{i,j=1}^{N} \Omega_{ij}^2 = 1.$$

Using (4) in the LVQ system, we can interprete (3) as a function $E(\{\mathbf{w}^j\}_{j=1}^{M}, \Omega)$ of, both, prototypes and parameterized relevance matrix. Standard optimization methods, including stochastic or batch gradient descent can be used to train the GMLVQ system from a given data set [6].

Typically, a feature-wise z-score transformation is performed based on the actual set of training examples [3,20]. Then, diagonal elements of Λ_{ii} after training can be interpreted as the relevance of single feature dimensions in the obtained classification, while off-diagonals reflect possible inter-dependencies between features [3,6,20].

3.2 Workflow

From any given data set $\mathbb{D}$, we generated 50 random splits of the P available samples into $Q \approx 0.9 \cdot P$ samples for the actual training and $(P-Q) \approx 0.1 \cdot P$ for validation. In every split, a z-score transformation was performed on the actual training set and the obtained mean values and standard deviations were also used to rescale the $P - Q$ valdiation samples accordingly.

After training, the validation performance was determined in terms of accuracies. In particular the Receiver Operator Characteristics (ROC) including the Area under the Curve (AUC) [11] were determined. This was achieved by introducing a varying margin in the determination of the winning protoype when classifying a novel sample, thus obtaining all possible combinations of *true positive* and *false positive rates* [6,11]. All results presented and discussed in Sect. 4, including ROC, diagonal relevances, and prototypes, were computed on average over the 50 training runs in a given experiment. In case of the ROC, we present the threshold averaged characteristics and its AUC [11], as well as the mean and standard deviations over the 50 individual results.

For GMLVQ training we employed batch gradient descent with step size control by waypoint-averaging [17], as available in the *No-nonsense GMLVQ toolbox* (version 2.3) [5]. We used the default parameter settings and performed 50 gradient steps in each training process.

In addition, RF classifiers were trained for comparison with GMLVQ in the same classification problems and setups. Specifically, we employed the built-in TreeBagger function of MATLAB (TM), version 2016a, with $n_{trees} = 50$ and other parameters set to their default values.

4 Results and Discussion

We focus on reporting results for the GMLVQ systems. In all experiments presented here, the outcomes of the RF classifiers were comparable or only slightly

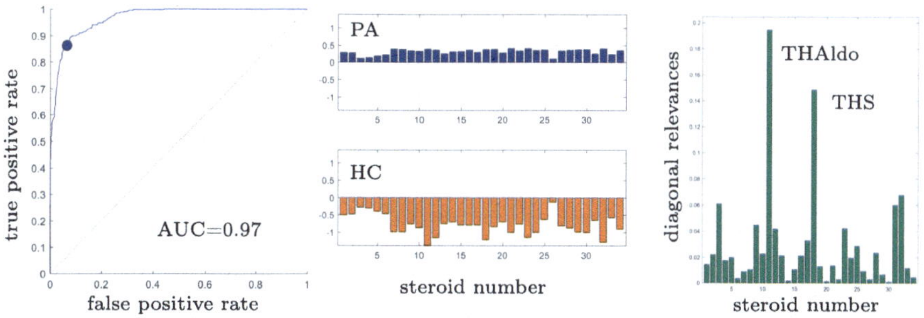

Fig. 2. Healthy Controls (HC) vs. Primary Aldosteronism (PA) patients. **Left panel:** threshold-averaged validation set ROC [11]. The filled circle marks the performance of the NPC. **Center panel:** prototypes in terms of z-score transformed feature values in the training set, averaged over 50 validation runs. **Right panel:** diagonal elements of relevance matrix Λ, averaged over 50 validation runs.

inferior. Note that also the *importances* [9] obtained in the RF are in agreement with the diagonal relevances of GMLVQ with respect to the three leading metabolites in all of the considered problems, see [18] for details.

4.1 Healthy Controls vs. Primary Aldosteronism

We first present results for the discrimination of HC (65 subjects) from patients presenting with PA (158 patients). The GMLVQ classifier achieved near perfect classification with an AUC of approximately 0.97 in the threshold averaged ROC. In the individual validation sets we also observed a mean AUC of 0.97 with standard deviation (std. dev.) 0.03.

While the obtained prototype configurations display overall higher excretion values in PA compared to HC, the relevance of steroids varies significantly within the panel, see Fig. 2. Inspection of the diagonal Λ_{ii} revealed that the classification relies to a large extent on the excretion values for THAldo (tetrahydroaldosterone, steroid number 11 in the panel)), and THS (Tetrahydro-11-deoxycortisol, number 19), see Fig. 2. In terms of relevance, the following metabolite is 18-OH-THA (18-Hydroxy-tetrahydro-11-dehydrocorticosterone, number 32).

Unsurprisingly, the main aldosterone metabolite THAldo plays the key role in the diagnosis of PA. Note that, here, virtually the same performance could be achieved (AUC = 0.96, std. dev. = 0.03) when using the three most relevant markers only, see [18].

4.2 Unilateral PA vs. Bilateral PA

The discrimination of UPA (88 patients) and BPA (70 patients) turned out a more challenging problem, see Fig. 3. The prototype configurations shown in the center panel suggest that BPA display systematically higher excretion of almost all steroids compared with UPA samples. However, GMLVQ achieved

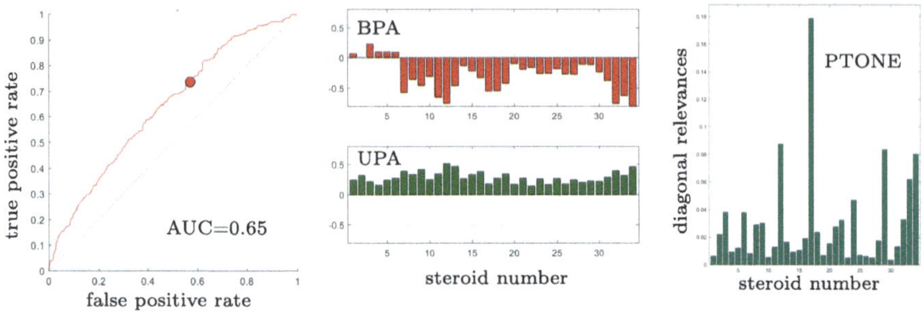

Fig. 3. Bilateral PA (BPA) vs. Unilateral PA (UPA) patients. Panel information analogous to Fig. 2.

only slightly better than random classification accuracy in the validation sets, with a mediocre AUC of approximately 0.65 in the threshold averaged ROC (mean AUC = 0.65 and std. dev. = 0.15). Here, the leading metabolite in terms of relevance turned out to be PTONE (Pregnanetriolone, number 17 in the panel), see also Fig. 3.

While the classification performance could be slightly improved by selecting sub-panels of only a few metabolites, the main result remains the same [18]. The classification accuracies were relatively poor indicating that a the general discrimination of BPA from UPA is challenging within the suggested approach.

It is important to note that the available cohort of UPA patients is relatively heterogeneous: As revealed by targeted gene sequencing, it comprises 34 samples with no identified somatic mutation in the tumor tissue, while in each of the remaining samples one of seven different driver mutations was found. Among these, the KCNJ5 mutation was the most frequent one, present in 31 of the 88 UPA patients.

This inspired our attempt to train a classifier for detecting the presence of the KCNJ5 motivation by means of the steroid profile alone, see following subsection.

4.3 KCNJ5 vs Non-KCNJ5 Samples

Motivated by the heterogeneity of the UPA cohort and the suboptimal performance described in the previous subsection, we also considered the discrimination of samples with identified KCNJ5 mutation (31 patients) from all other PA samples, i.e. from UPA patients with no or a different mutation combined with all BPA patients (127 samples in total).

In this *KCNJ5 vs. non-KCNJ5* classification problem, GMLVQ achieved relatively good validation performance with an average AUC of 0.83 (std. dev. = 0.16), see Fig. 4. The by far most relevant steroid metabolite was 18-Oxo-THF (18-Oxo-tetrahydrocortisol, number 34). Note that this marker also played a significant yet less pronounced role in the attempted discrimination of *UPA vs. BPA* discussed in the previous subsection.

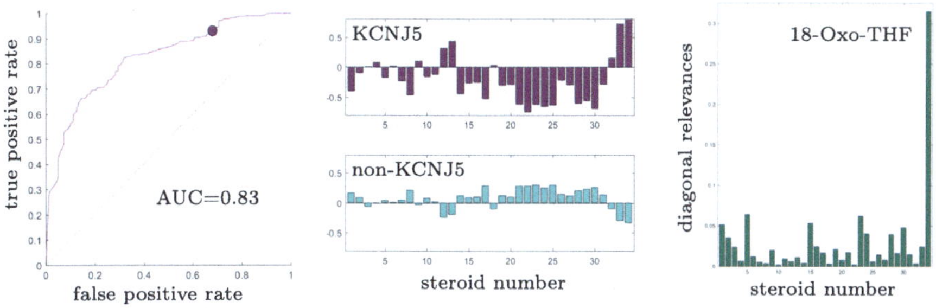

Fig. 4. UPA samples with identified KJCN5 mutation vs. all other UPA and BPA patients. Panel information analogous to Fig. 2.

Here, restricting the data set to the three most relevant metabolites, i.e. 18-Oxo-THF (34), THF (21) and 18-OH-F (33), leads to a nominal improvement of the performance with an AUC of 0.85 (std. dev. = 0.16).

KCNJ5 samples appear to display a pronounced pattern of steroid excretions in comparison with the UPA profiles in absence of this particular mutation and in BPA. This is evidenced by the protoype configurations shown in Fig. 4.

5 Conclusion and Outlook

Our machine learning analysis shows that USM can reliably differentiate PA patients from healthy controls. More importantly, the non-invasive method can identify KCNJ5-mutated UPA with very good sensitivity and specificity. In clinical practice, this could bypass the invasive and involved adrenal vein sampling for patients with unilateral adenomas. KCNJ5 driven UPA could be identified directly from urinary steroid profiles, facilitating subtype-specific treatment [18]. Further studies should be pursued in order to validate the usefulness of the suggested method as a practical diagnostic test.

Future improvements of the machine learning approach will address the deeper analysis of feature relevances, e.g. along the lines suggested in [14]. On the one hand, the *Iterated Relevance Matrix Analysis* (IRMA) could facilitate the construction of superior classifiers for the identification of KCNJ5 driven PA. On the other hand, the improved interpretation of steroid relevances beyond the dominance of 18-Oxo-THF might shed light on the disease mechanisms behind this subtype of Primary Aldosterone.

Ultimately, the more detailed study of other mutation driven UPA hinges on the availability of larger data sets.

Acknowledgments. Alessandro Prete receives support from the National Institute for Health and Care Research (NIHR) Birmingham Biomedical Research Centre at the University Hospitals Birmingham NHS Foundation Trust and the University of Birmingham (Grant Reference Number NIHR203326).

References

1. Alahmadi, H.H., et al.: Classifying cognitive profiles using machine learning with privileged information in mild cognitive impairment. Front. Comput. Neurosci. **10**, 117 (2016)
2. Arlt, W., et al.: Urine steroid metabolomics as a biomarker tool for detecting malignancy in adrenal tumors. J. Clin. Endocrinol. Metab. **96**, 3775–3784 (2011)
3. Biehl, M., Hammer, B., Villmann, T.: Prototype-based models in machine learning. Wiley Interdiscip. Rev. Cogn. Sci. **7**, 92–111 (2016)
4. Biehl, M., et al.: Matrix relevance LVQ in steroid metabolomics based classification of adrenal tumors. In: Verleysen, M. (ed.) 20th European Symposium on Artificial Neural Networks (ESANN 2012), pp. 423–428. D-side Publishing (2012)
5. Biehl, M.: A collection of no-nonsense GMLVQ code. University of Groningen, The Netherlands (2016). https://www.cs.rug.nl/~biehl/gmlvq
6. Biehl, M.: The Shallow and the deep: a biased introduction to neural networks and old school machine learning. University of Groningen Press, 290 p, open access (2023)
7. Bittrich, S., Kaden, M., Leberecht, C., Kaiser, F., Villmann, T., Labudde, D.: Application of an interpretable classification model on early folding residues during protein folding. BioData Mining **12**, 1–16 (2019)
8. Bohnsack, K.S., Kaden, M., Voigt, J., Villmann, T.: Efficient classification learning of biochemical structured data by means of relevance weighting for sensoric response features. In: Verleysen, M. (ed.) Proceedings of 27th European Symposium on Artificial Neural Networks ESANN 2022. i6doc.com (2022)
9. Breiman, L.: Random forests. Mach. Learn. **45**, 5–32 (2001)
10. Chortis, V., et al.: Urine steroid metabolomics as a novel tool for detection of recurrent adrenocortical carcinoma. J. Clin. Endocrinol. Metab. **105**(3), e307–e318 (2019)
11. Fawcett, T.: An introduction to ROC analysis. Pattern Recogn. Lett. **27**, 861–874 (2006)
12. Hammer, B., Villmann, T.: Generalized relevance learning vector quantization. Neural Netw. **15**(8–9), 1059–1068 (2002)
13. Kohonen, T.: Self-Organizing Maps, 2nd edn. Springer, Heidelberg (1997). https://doi.org/10.1007/978-3-642-56927-2
14. Lövdal, S., Biehl, M.: Iterated relevance matrix analysis (IRMA) for the identification of class-discriminative subspaces. Neurocomputing **577**, 127367 (2024)
15. Moolla, A., et al.: Accurate non-invasive diagnosis and staging of non-alcoholic fatty liver disease using the urinary steroid metabolome. Aliment. Pharmacol. Ther. **51**(11), 1188–1197 (2020)
16. Nova, D., Estévez, P.: A review of learning vector quantization classifiers. Neural Comput. Appl. **25**(3–4), 511–524 (2014)
17. Papari, G., Bunte, K., Biehl, M.: Waypoint averaging and step size control in learning by gradient descent (technical report). In: Schleif, F.M., Villmann, T. (eds.) MIWOCI 2011, Mittweida Workshop on Computational Intelligence. Machine Learning Reports, vol. MLR-2011-06, pp. 16–26. University of Bielefeld (2011)
18. Prete, A., et al.: Urine steroid metabolomics as a diagnostic tool in primary aldosteronism. J. Steroid Biochem. Mol. Biol. **237**, 106445 (2023)
19. Sato, A., Yamada, K.: Generalized learning vector quantization. In: Tesauro, G., Touretzky, D., Leen, T. (eds.) Advances in Neural Information Processing Systems, vol. 7, pp. 423–429 (1995)

20. Schneider, P., Biehl, M., Hammer, B.: Adaptive relevance matrices in learning vector quantization. Neural Comput. **21**, 3532–3561 (2009)
21. Storbeck, K.H., et al.: Steroid metabolome analysis in disorders of adrenal steroid biosynthesis and metabolism. Endocr. Rev. **40**(6), 1605–1625 (2019)
22. Taylor, A.E., et al.: Urine steroid metabolomics for the differential diagnosis of adrenal incidentalomas in the eurine-act study: a prospective test validation study. Lancet Diab. Endocrinol. **8**(9), 773–781 (2020)
23. van Veen, R., et al.: An application of generalized matrix learning vector quantization in neuroimaging. Comput. Methods Programs Biomed. **197**, 105708 (2020)

The Beauty of Prototype Based Learning

Peter Tino(✉)

School of Computer Science, University of Birmingham, Birmingham, UK
p.tino@bham.ac.uk

Abstract. There is a long history of formulation of machine learning models that have at their core "representative prototypes" living in the data space (examples include variants of Self Organizing Maps in the unsupervised setting or variants of Learning Vector Quantization in the supervised one). The prototype based formulations make such models intuitive and naturally amenable to interpretation. I will present a series of generalizations and extensions of this idea, arguing that prototype based learning still deserves attention in the current era of deep learning. In particular, I will present extensions of prototype based learning to model spaces and Riemannian manifolds. I will also show how prototype based models can be naturally extended to the setting of ordinal regression and learning with privileged information.

T. Villmann et al. (Eds.): WSOM+2024, LNNS 1087, p. 106, 2024.
https://doi.org/10.1007/978-3-031-67159-3_12

Setting Vector Quantizer Resolution via Density Estimation Theory

Josh Taylor(✉) and Stella Offner

The University of Texas at Austin, Austin, TX, USA
joshtaylor@utexas.edu

Abstract. We introduce a framework for selecting the number of codebook vectors in a vector quantizer based on local characteristics of the data density, the degree to which the process of VQ distorts the representation of this density, and the theoretical efficiency of estimators of these densities. In our analysis, L^2 theory from kernel density estimation relates the number of VQ prototypes to observed sample size, dimension, and complexity, all of which intuitively influence codebook sizing.

Keywords: Vector Quantization · Variable Kernel Density Estimation

1 Introduction

Vector Quantization (VQ, [11]) is a technique for data encoding and compression used for a variety of pattern matching tasks in many domains. Given a d-dimensional sample of data $X = \{x_s \in \mathbb{R}^d\}_{s=1}^N$ drawn from density $f(x)$, VQ algorithms learn a set of *prototype* or *codebook* vectors $W = \{w_i \in \mathbb{R}^d\}_{i=1}^M$, where typically $M << N$, to approximate $f(x)$. The quantization function $Q(x)$ represents x by its **B**est **M**atching **U**nit (BMU) in W, which is the element of W with minimum r-th order distortion (distance to x, also called quantization error). Voluminous VQ research (see [12] for a thorough overview) has focused primarily on theoretical considerations and implications of VQ design for either *a fixed or asymptotically increasing number of prototypes.* That is, given M, much is known about optimal W, the behavior of the resulting quantization error, the geometries of the resulting Voronoi partition, etc. By contrast, this body of research has very little to say about specifying M in practical settings: VQ theorists often assume M "large enough" to invoke limiting behaviors in proofs; VQ practitioners may select M to meet constraints such as ensuring a maximum quantization error, or size of the encoded signal. While situational constraints may be used to set M in some domains, we are interested in VQ as a tool to build *prototype-based models in machine learning* [2], which include, e.g., the Self-Organizing Map (SOM, [14]), Neural Gas (NG, [20]), Learning Vector

Authors acknowledge partial support from NSF AAG 2107942 and seed funding from NSF's Institute for Foundations in Machine Learning.

T. Villmann et al. (Eds.): WSOM+2024, LNNS 1087, pp. 107–117, 2024.
https://doi.org/10.1007/978-3-031-67159-3_13

Quantization (LVQ, [14]), and even K-means [19]. These algorithms are regularly applied to domains where no hard constraints guide the selection of M.

Various rules-of-thumb and algorithmic enhancements have appeared over the years to address this issue. In what follows, we present a framework for choosing M based on concepts from density estimation theory. In Sect. 3 we review methods for selecting M in practice; Sect. 4 presents a review of the statistical theory of kernel density estimation, curated for our current needs; Sect. 5 outlines our density-based framework, which is exercised on synthetic data in Sect. 6.

2 VQ Theory

Optimal VQ design seeks the best codebook W to represent $f(x)$, as measured through minimum expected distortion, also called Quantization Error: [10]:

$$E[D_r] = \int ||x - w_{b(x)}||^r \, f(x) \, dx \qquad (1) \qquad\qquad W = \underset{\{w_i\}_{i=1}^M}{\arg\min} \, E[D] \qquad (2)$$

While (1) has motivated the study of many theoretical aspects of VQs over the last half-century, we briefly present two theoretical results first presented by Zador [26] germane to this work. Both involve a crucial quantity $\alpha = \frac{d}{d+r}$, which we call the *magnification exponent.* **Z1:** Evaluated at its theoretical optimal W (2), distortion (1) tends to 0 at rate $\propto ||f||_\alpha M^{-\frac{r}{d}}$ as $M \to \infty$. Here, $||f||_p$ denotes the L^p function norm. **Z2:** The prototypes $W = \{w_i\}_{i=1}^M$ (2) are joint minimizers of (1) if and only if the point density of prototypes, which we denote $g(w)$, obeys the following power law with respect to the data density f:

$$g(w) = cf(w)^\alpha, \quad c = 1/\int f(w)^\alpha \, dw \qquad (3)$$

The consequence of **Z1** is that an optimal VQ with larger M *always* results in lower quantization error, which may suggest over-specifying M. Since prototype-based ML models can lose pertinence or efficacy as M increases and the VQ mapping tends toward the identity function, we have reason to prefer a more parsimonious approach. One implication of **Z2** is that prototypes resulting from an optimal VQ can be thought of as a sample of size M from $g = cf^\alpha$. As defined, $\alpha < 1$ *always* with $\lim_{d \to \infty} \alpha = 1$. Consequently, the density surface defined by g magnifies parts of input space which have low measure according to f, and usually ensures g a smoother density than f, which will be important in Sect. 5 In practice, we are not given $f(x)$ but rather X, a sample of size N assumed to be drawn from it, so the expectation of (1) is replaced by its empirical mean. Unfortunately, this guarantees that minimization of (1) is NP hard. Nevertheless, a collection of practical algorithms [8,16,17,19] have been developed by both engineers and statisticians to fit VQ models to observed data X. These classical algorithms above are all based on *competitive* learning, indicating that the prototypes "compete" to represent (quantize) each x_s. Neural VQs, such as the SOM [14] and Neural Gas [20], add a *cooperative* element to their iterative update rules where, at each learning step, prototype updates are influenced both by their own receptive fields (like classical VQ) and by the RFs of neighboring prototypes.

However, cooperation is known [6,14] to impart a new magnification exponent α^* on the resulting prototype density g. While several works have derived theoretical expressions for α^* in restricted settings, no form exists for a neural VQ with arbitrary but fixed M in general dimension d, although it is "certainly less than unity" [14, Chapter 3]. While we will not attempt to add any further insight to α^* in this work, we have highlighted the different magnification behaviors of neural VQs to stress that the results of Sect. 5 will differ for them.

3 Existing Methods to Select M

In practice, VQ sizing is often done via trial-and-error, where M is specified along a grid, VQ codebooks of size M are built, some measure of fit (e.g., quantization error) computed, and the process repeated until a "good" codebook size is determined (or time/computational constraints are met); we stress the absence of any universally acceptable definition of "good." Various rules of thumb abound, mostly in practical handbooks (e.g., [15]) and documentation for VQ algorithms, with a wide range of suggested starting values for M. Letting $M = \mathcal{O}(\sqrt{(}N))$ is a popular starting point for grid searching, which we note has analogs in Silverman's rule of thumb for selecting the resolution in KDEs [23]. Less ad-hoc methods to select M exist, but most require that multiple VQs be fit [7], shift the burden of specifying M to that of specifying alternate (possibly more obtuse) parameters [9] or require additional information such as labeled data [24]. We prefer a more self-supervised and formal approach. To do so, in the next section we invoke theory from kernel density estimation where researchers have battled the dilemma of optimal data resolution for over half a century.

4 Variable Kernel Density Estimation

The advantage of VQ, and the crux of our methodology, depends on its density matching properties governed by the power law (3). As such, we pause now to introduce relevant parts of density estimation theory of use in this work. Specifically, we focus on variable KDEs, as they are better suited to higher-dimensional settings [25], harkening the space filling advantage of VQs [18].

Kernel density estimation utilizes smoothing kernel functions K centered at observed data $X = \{x_s\}_{s=1}^N$ to approximate the density $f(x)$ from which X is assumed to be arise. Variable kernel density estimators (vKDEs) prescribe different orientations and levels smoothing amounts across $\mathbb{R}^d$, whereas *fixed* KDEs use the same smoothing parameters everywhere. Fixed KDEs are more commonly employed, mostly because their construction is simpler but vKDEs are capable of better performance in high-dimensional settings. We note that most of the theory below is summarized from [25], but we borrow some notation from [4] for mathematical brevity. A multivariate vKDE has the form

$$f_y \approx \tilde{f}_y = \frac{1}{N}\sum_{s=1}^{N} K_{H_y}(y - x_s), \quad K_{H_y} = |H_y|^{-1}K(H_y^{-1}y), \tag{4}$$

where $y \in \mathbb{R}^d$ is an arbitrary point and subscripts y denote function evaluation at y. The kernel parameter H_y is called the *bandwidth* matrix prescribing the scale and rotation of smoothing K asserts amongst the sample points to produce the estimate $\hat{f}$. In contrast with fixed estimators, H_y varies with y.

H_y is the most important tuneable parameter of (4); we re-parameterize it as $H_y = h_y A_y$, where $h_y > 0$ is a scalar controlling the kernel's size and $A_y : |A_y| = 1$ is a $d \times d$ rotation matrix controlling its elliptical shape. To achieve consistency, we require that $h_y \to 0$ as $N \to \infty$. The exact functional form of K is less crucial to the success of (4) [23], but K is generally constrained to be a probability density in its own right satisfying the following moment conditions:

$$(i) \int K(y)\,dy = 1; \quad (ii) \int yK(y)\,dy = 0; \quad (iii) \int yy^T K(y)\,dy = I_d \tag{5}$$

where I_d is the d-dimensional identity matrix. For computational simplicity, K is often assumed to be the pdf of standard multivariate Gaussian distribution.

4.1 L^2–Error for vKDEs

We now give an overview of squared error analysis of (4), which is crucial to the theory of KDEs and motivates the methodology presented in Sect. 5. Detailed derivations of these results are found in, e.g., [25], [23, Chapter 6] and [4]. At an arbitrary evaluation point y, pointwise squared error of (4) with bandwidth parameter $H_y = h_y A_y$ is given by $(\hat{f}_y - f_y)^2$. Because $\hat{f}_y$ is a random variable (via its dependence on our sample X), it is more appropriate to consider pointwise Mean Squared Error $MSE\left[\hat{f}_{y|X}\right] = E_X\left[(\hat{f}_{y|X} - f_y)^2\right]$ where we have used the notation $f_{y|X}$ to remind us that the $\hat{f}_y$ is a function of the random sample X. Pointwise *bandwidth selection* in kernel density estimation seeks H_y as the arg min of $MSE\left[\hat{f}_y\right]$, which is often intractable (even if true f is known). To circumvent this, *asymptotic* error analysis replaces f with its Taylor series expansion in a window of shape/size H_y about y. Ignoring order-3 terms and above in the series results in the following Asymptotic Variance (AV) and Squared Bias (ASB) decomposition (from [23], with notation from [4]):

$$AMSE[\hat{f}_y] = AV[\hat{f}_y] + ASB[\hat{f}_y] \tag{6}$$

$$AV[\hat{f}_y] = \frac{f_y R[K]}{h_y^d N} \tag{7}$$

$$ASB[\hat{f}_y] = h_y^{2p_y}\beta_y \tag{8}$$

$$\beta_y = \left[\frac{1}{p_y!}\mu_{p_y}^T(K)\, A_y^{\otimes p_y}\, \mathcal{D}^{\otimes p_y} f_y\right]^2 \tag{9}$$

In the above, $R[K] = \int K(z)^2\, dz$ denotes the statistical *roughness* of K and $p_y \in \{2, 4\}$ is the *order* of the estimate, based on the curvature of f_y ($p = 2$ when the Hessian $\nabla^2 f_y$ is positive or negative definite, $p = 4$ when it is indefinite). μ_{p_y} is the vectorized p_y-th moment of the kernel, defined as $\mu_p(K) = \int z^{\otimes p} K(z)\, dz$ where $z^{\otimes p}$ is the p-th fold Kronecker product of the vector $z \in \mathbb{R}^d$ with itself (a d^p length vector, after vectorization). $A_y^{\otimes p_y}$ is the p_y-th Kronecker product of A with itself (a $d^{p_y} \times d^{p_y}$ matrix), and $\mathcal{D}^{\otimes p_y} = \frac{\partial f}{\partial y^{\otimes p_y}}$ is the vectorized p_y-th fold Kronecker power of the differential operator applied to f_y whose d^{p_y} elements contain all mixed partials of f up to order p_y. More explanation of vectorized higher-order derivatives can be found in [5].

Pointwise bandwidth selection minimizes $AMSE[\hat{f}_y]$ over h_y and A_y, which is separable in these parameters as the latter enters only via β_y. [25] describes this minimization over A_y in great detail as its procedure various based on p_y; for brevity we reproduce the optimal bandwidth results below. Letting A_y^* be the minimizer of β_y and β_y^* its minimum value, the optimal bandwidth scale h_y and resulting AMSE is given by

$$h_y^* = \left(\frac{d\, f_y\, R[K]}{2p_y\, \beta_y^* N} \right)^{1/(d+2p_y)} \tag{10}$$

$$AMSE^* \left[\hat{f}_y \right] = \epsilon[\hat{f}_y]\, N^{-2p_y/(d+2p_y)} \tag{11}$$

$$\text{where } \epsilon[\hat{f}_y] = \left(1 + \frac{d}{2p_y} \right) (f_y R[K])^{2p_y/(d+2p_y)} \left(\frac{2p_y}{d} \beta_y^* \right)^{d/(d+2p_y)} .$$

(11) tells us that vKDE error decreases with sample size N at a rate $\mathcal{O}(N^{-2p_y/(d+2p_y)})$, which is faster than fixed KDEs most everywhere [25]. We harness this improvement in the methodology for sizing VQ codebooks described below.

5 Methodology

5.1 Motivation

Our goal is to deduce a relationship between the number of VQ prototypes M, the complexity of our data X, and the observed sample size N. This is motivated by the assumption that encoding a large amount of complicated data demands more prototypes – but how many more? The answer depends on how one characterizes the loose notion of data complexity, which we formalize here.

Zador's results summarized in Sect. 2 reveal that the M prototypes of an optimal VQ follow the density $g = cf^{\alpha}$. Thus, fitting a VQ in practice with sample X of size N results in a prototype sample W (from density g) of size M. Density estimation theorists have spent decades formally characterizing how (i.e., the process of kernel bandwidth selection), and to what degree of accuracy (i.e., the resulting minimal AMSE), one can infer the underlying density from which a sample arises. Deeper scrutiny of their main conclusions, as encapsulated in the terms of the optimal AMSE (11), reveal several analogues to data considerations germane to VQ practitioners when specifying M, such as sample size, dimension, and data complexity (terms involving f, $\mathcal{D}f$, etc.). **Our observation is that these M prototypes, if optimally derived, should represent *their* density g *as well as* the data X represent *their* density f.** That is, we propose selecting M to best represent g, as measured via AMSE, modulo the constraints on this process inherent in the observed sample X from which W arises.

5.2 An Equivalent Sample Size Analysis for M

Statisticians compare estimators in terms of their relative *efficiency*, as measured by their error. At evaluation point y,, we let $\tilde{f}_y$ and $\tilde{g}_y$ denote the theoretically optimal variable density estimates of f_y and g_y based on samples X and W, respectively. That is, $\tilde{f}_y$ and $\tilde{g}$ have functional form given by (4) with sample sizes N and M, respectively. The *efficiency* of $\tilde{g}$ to $\tilde{f}$ can be characterized by quantity η, which reports the ratio of their minimal errors:

$$\eta = \frac{AMSE^*[\tilde{g}]}{AMSE^*[\tilde{f}]} = \frac{\epsilon[\tilde{g}_y]\, M^{-2q_y/(d+2q_y)}}{\epsilon[\tilde{f}_y]\, N^{-2p_y/(d+2p_y)}} \tag{12}$$

where, for clarity, we have let q_y denote the order of g_y (the analog of p_y for f_y). For given values of its RHS (including M and N), η conveys how much better ($\eta \leq 1$) or worse ($\eta > 1$) $\tilde{g}_y$ is at estimating g_y than $\tilde{f}_y$ is at estimating f_y; we now have a functional form relating data complexity (the admittedly complex ratio $\epsilon[\tilde{g}_y]/\epsilon[\tilde{f}_y]$), the observed sample size N, and the number of prototypes M. Prescribing an acceptable efficiency η allows us to solve for the number of prototypes required to achieve such efficiency at y, which is known as *equivalent sample size analysis* in density estimation theory:

$$M_y^* = \left(1/\eta \times \epsilon[\tilde{g}_y]/\epsilon[\tilde{f}_y] \times N^{2p_y/(d+2p_y)}\right)^{(d+2q_y)/(2q_y)}. \tag{13}$$

Taking the expectation of M_y^* over our sample gives an estimate of the aggregate number of prototypes required to achieve efficiency η overall:

$$M^* = \frac{1}{N}\sum_{s=1}^{N} M_{x_s}^*. \tag{14}$$

η is the only tuneable parameter in our methodology and has a direct, interpretable meaning; e.g., specifying $\eta = 2$ results in a number of prototypes which represent $g = cf^\alpha$ *twice* as badly as f could be represented with its N observations. Thus, a practitioner only needs to choose their relative error tolerance prior to constructing a VQ codebook.

5.3 Practical Considerations

Evaluating M^* in practice is not trivial for various reasons and we pause now to address computational roadblocks. Assuming f is known (for now), Zador's conclusions relate the prototype density $g = cf^\alpha$ but we still must compute its normalizing constant c. Empirical estimation of $c = E_X[f(X)^{\alpha-1}]$ is an option, but this will suffer from the "curse of dimensionality" unless increasingly large sample sizes are available. Instead, for now, we propose restricting f to the class of Gaussian Mixture Models ($f = \sum_k w_k \phi(\mu_k, \Sigma_k)$) and approximating true g by the straightforward method of [1]. When true f is not known (most practical settings), we propose using a preliminary *pilot* estimate $\hat{f}_0$ directly from X using Gaussian kernels, which is common among modern kernel smoothing methods [4]. Finally, because the components of AMSE involve terms whose size grows exponentially with data dimension d, computation of (11) demands the careful algorithms of [5] for feasibility in arbitrary dimension.

6 Experiments and Discussion

We selected four Gaussian mixtures f from density estimation literature [3] to showcase AMSE-modulated M^*. These mixtures, abbreviated Mix. $\{8, 9, 11, 12\}$, have varying number of components and component-wise covariance structures resulting in different complexities. The mixtures in the literature are all $2d$; because we would like to showcase M^* across a range of d we have expanded their dimensionality by replicating each component-wise covariance across the range $d \in \{2, 4, 6, 8, 10, 20, 40, 60, 80, 100\}$. Visuals of these mixtures in $2d$ are given below, where colors correspond to the various mixture components.

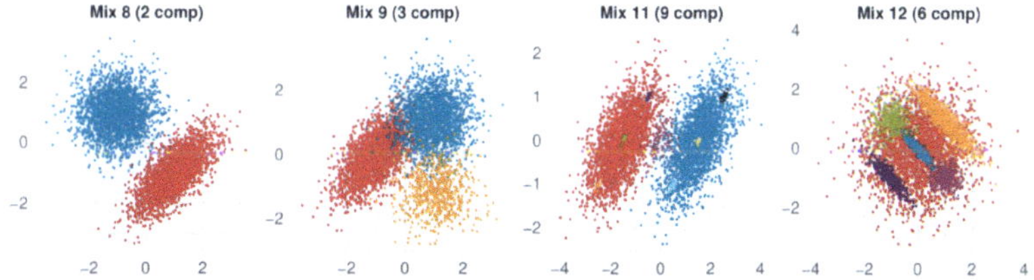

Each mixture and dimension listed above corresponds to a known data density f from which we estimate the prototype density g according to [1]. From each f we sample $X \sim f$ of size $N = 10,000$ and compute the number of prototypes M^* according to (14). To assess the suitability of M^* we view it in relation to [1] an information-theoretic characterization of data complexity, and [2] the Mean Quantization Error resulting from K-means across a large range of M.

For measure [1], we compute the overall mean $\bar{\mu}$) and covariance $\bar{\Sigma}$ for each $f(m, d)$. Because the multivariate Gaussian density has maximum entropy for continuous unbounded data with a given first and second moment, the KL-Divergence from $N(\bar{\mu}, \bar{\Sigma})$ to $f(m, d)$ represents how far, in an information-theoretic sense, $f(m, d)$ is from maximum entropy; i.e., it is *one* measure of data complexity that is unitless (comparable across varying d). We posit that higher-entropy mixtures should require more prototypes, which is generally confirmed in Fig. 1. For each mixture KL increases with d, which is not unexpected as it is known that higher-dimensional data tend to form "hubs" (clumps) in space more strongly [21]. While M^* generally decreases with KL it is not a completely monotonic trend which is likely due to several factors: (a) we have empirically estimated both KL and M^* from our sample of size $N = 10,000$ so both are subject to typical statistical estimation errors; (b) likely more influential is the suitability of KL as the only oracle for data complexity.

In practice, it is common to build several VQs for the same data by varying M and observing how the quantization error (distortion) behaves in response. So-called knee-finding algorithms (e.g., [22]) are often used to help guide the eye to points of regime switching along the QE vs. M curve. These curves, resulting from fitting K-means for M in the range $[10, 9000]$ are shown in Fig. 2 for our two most visually complicated mixtures (11 and 12). In each case, the M^* selected via (14) is shown on the curve (salmon points), along with the curve's knee (dark blue points) identified by [22].

Comparing the two, we see in Fig. 1 that M^* appears a conservative recommendation for the number of prototypes, at least when we demand $\eta = 1$. We also note the trend that as d increases $M^* \to M_{knee}$, suggesting M^* may signal regime changes along the QE curve *without* the burden of constructing many different quantizers, at least in higher-d cases.

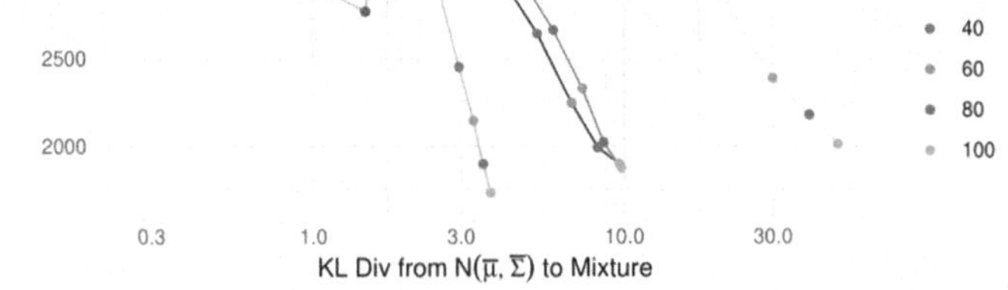

Fig. 1. M^* from (14) with $\eta = 1$ vs. KL-Divergence from a $N(\bar{\mu}, \bar{\Sigma})$ density to each mixture of noted dimension. A larger KL divergence indicates the mixture is less complicated, from an Information Theory perspective, relative to its overall scale.

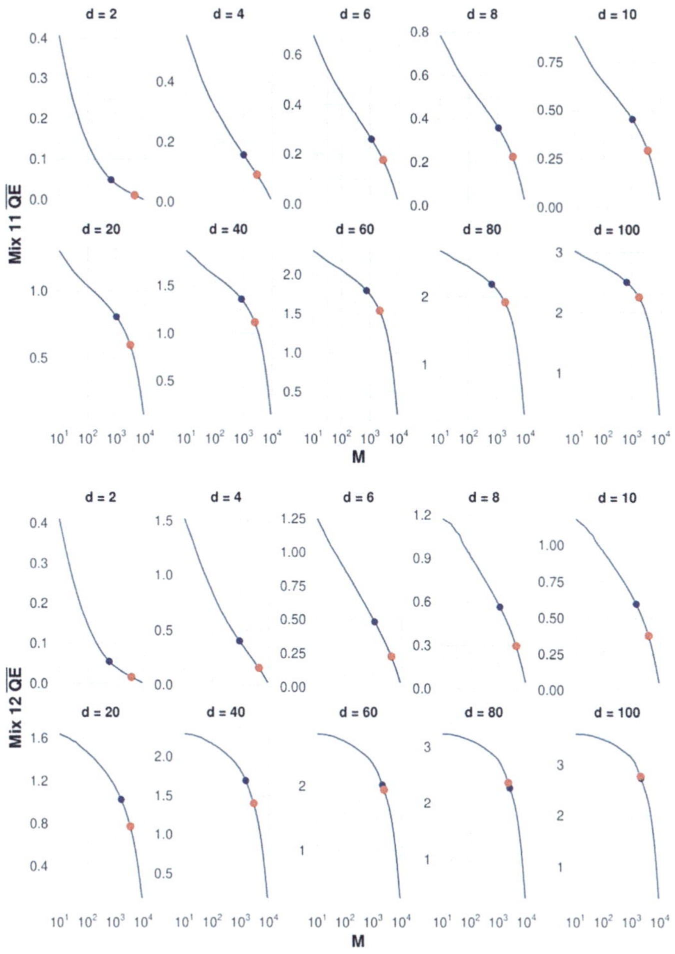

Fig. 2. Mean quantization error for Mixtures 11 and 12 (dark lines) produced by K-means along a range of M. The salmon points indicate the M^* selected via our method, and the dark blue points are the result of the knee-finding Kneedle algorithm [22].

7 Conclusions and Further Work

We have presented a framework for selecting the number of prototypes (codebook vectors) in a vector quantizer modulated by squared-error theory of kernel density estimation. The link between these two relies on a subset of VQ theory defining the prototype density of an optimal quantizer. Crucially, our equivalent sample size analysis, common in analyses of statistical estimators, replaces ad-hoc or rule-of-thumb methods for sizing codebooks with a formalism whose only parameter η is a meaningful measure of relative estimator efficiency. Specification of M^* is completely *a-priori*, alleviating the need to fit multiple vector quantizer(s). Of further benefit, our methodology can be inverted to reveal the (relative) impact of a given codebook size on the ability for resulting prototypes to to represent *their* density (and, in turn, the data density).

For this introductory work we have restricted experimentation to synthetic cases where the data density f is known; in the future we will remove this restriction by prescribing a pilot estimation scheme such as in [4] to select M in more general settings. Inspired by [13], we also believe uniformly demanding $\eta = 1$ (as done in our initial experiments) may err on a conservative estimate M^*; further experiments with a range of η efficiencies are planned.

References

1. Ajgl, J., Šimandl, M., Duník, J.: Approximation of powers of gaussian mixtures. In: 18th International Conference on Information Fusion, pp. 878–885. IEEE (2015)
2. Biehl, M., Hammer, B., Villmann, T.: Prototype-based models in machine learning. Wiley Interdisc. Rev. Cogn. Sci. **7**(2), 92–111 (2016)
3. Chacón, J.E.: Data-driven choice of the smoothing parametrization for kernel density estimators. Can. J. Stat. **37**(2), 249–265 (2009)
4. Chacón, J.E., Duong, T.: Multivariate kernel smoothing and its applications. Chapman and Hall/CRC (2018)
5. Chacón, J.E., Duong, T.: Higher order differential analysis with vectorized derivatives. arXiv preprint arXiv:2011.01833 (2020)
6. Cottrell, M., Fort, J.C., Pagès, G.: Theoretical aspects of the SOM algorithm. Neurocomputing **21**(1–3), 119–138 (1998)
7. De Bodt, E., Cottrell, M., Verleysen, M.: Statistical tools to assess the reliability of self-organizing maps. Neural Netw. **15**(8–9), 967–978 (2002)
8. Forgey, E.: Cluster analysis of multivariate data: efficiency vs. interpretability of classification. Biometrics **21**(3), 768–769 (1965)
9. Fritzke, B.: A growing neural gas network learns topologies. In: Advances in Neural Information Processing Systems, vol. 7 (1994)
10. Gersho, A.: Asymptotically optimal block quantization. IEEE Trans. Inf. Theory **25**(4), 373–380 (1979)
11. Gersho, A., Gray, R.M.: Vector Quantization and Signal Compression, vol. 159. Springer, New York (2012)
12. Gray, R.M., Neuhoff, D.L.: Quantization. IEEE Trans. Inf. Theory **44**(6), 2325–2383 (1998)
13. Gray, R.M., Olshen, R.A.: Vector quantization and density estimation. In: Proceedings. Compression and Complexity of SEQUENCES 1997 (Cat. No. 97TB100171), pp. 172–193. IEEE (1997)
14. Kohonen, T.: Self-Organizing Maps, vol. 30. Springer, Heidelberg (2012)
15. Kohonen, T., et al.: Matlab implementations and applications of the self-organizing map. Unigrafia Oy, Helsinki, Finland **177** (2014)
16. Linde, Y., Buzo, A., Gray, R.: An algorithm for vector quantizer design. IEEE Trans. Commun. **28**(1), 84–95 (1980)
17. Lloyd, S.: Least squares quantization in PCM. IEEE Trans. Inf. Theory **28**(2), 129–137 (1982)
18. Lookabaugh, T.D., Gray, R.M.: High-resolution quantization theory and the vector quantizer advantage. IEEE Trans. Inf. Theory **35**(5), 1020–1033 (1989)
19. MacQueen, J.: Classification and analysis of multivariate observations. In: 5th Berkeley Symp. Math. Statist. Probability, pp. 281–297 (1967)

20. Martinetz, T.M., Schulten, K.J.: A "neural gas" network learns topologies. In: Kohonen, T., Mäkisara, K., Simula, O., Kangas, J. (eds.) Proceedings of the International Conference on Artificial Neural Networks 1991 (Espoo, Finland), pp. 397–402. Amsterdam; New York: North-Holland (1991)
21. Radovanovic, M., Nanopoulos, A., Ivanovic, M.: Hubs in space: popular nearest neighbors in high-dimensional data. J. Mach. Learn. Res. **11**(sept), 2487–2531 (2010)
22. Satopaa, V., Albrecht, J., Irwin, D., Raghavan, B.: Finding a "kneedle" in a haystack: Detecting knee points in system behavior. In: 2011 31st International Conference on Distributed Computing Systems Workshops, pp. 166–171. IEEE (2011)
23. Scott, D.W.: Multivariate Density Estimation: Theory, Practice, and Visualization. Wiley, Hoboken (2015)
24. Silva, L.A., de Vasconcelos, B.P., Del-Moral-Hernandez, E.: A model to estimate the self-organizing maps grid dimension for prototype generation. Intell. Data Anal. **25**(2), 321–338 (2021)
25. Terrell, G.R., Scott, D.W.: Variable kernel density estimation. Ann. Stat. 1236–1265 (1992)
26. Zador, P.: Asymptotic quantization error of continuous signals and the quantization dimension. IEEE Trans. Inf. Theory **28**(2), 139–149 (1982)

Practical Approaches to Approximate Dominant Eigenvalues in Large Matrices

Frank-Michael Schleif(✉)

Department of Computer Science and Business Information Systems, Center for Artificial Intelligence and Robotics UAS Würzburg-Schweinfurt, Würzburg, Germany
frank-michael.schleif@thws.de

Abstract. In numerous machine learning algorithms, the computation of eigenvectors and eigenvalues plays a pivotal role in optimizing models. Depending on the specific use case, it is frequently adequate to compute only a subset of dominant eigenvectors or utilize estimations. Handling this task for large matrices poses a challenge, as standard machine learning packages often lack suitable implementations. We explores various techniques for approximating dominant eigenvectors in the context of potentially large symmetric, real-valued matrices and offer an overview of established methods, analyzing their potentials and limitations, including implementation details.

Keywords: eigenvalue approximation · large scale · Nyström · Gershgorin · power iteration · *von Mises* · Lanczos

1 Introduction

Eigenvalues and eigenvectors are crucial elements in various machine learning approaches, with applications in optimization theory, information encoding, data representation and alike [8,9,19]. Standard machine learning frameworks like scikit-learn/numpy and other provide dedicated methods for eigenvalue decomposition of a given matrix, typically using the Lanczos iteration (see e.g. [1,14] for more recent work) as the most prominent method. The core work is in general done by highly optimized industrial libraries for numerical problems such as lapack [1]. These methods are helpful if one has to decompose a full matrix or needs many of the underlying eigenvalues and eigenvectors. However, scenarios like handling large matrices, finding only some dominating eigenvectors, checking for positive (semi) definiteness - p(s)d or implementing algorithms on low cost devices have particular requirements. This motivates the usage of different eigenvalue estimators or even individual more straight forward implementations. In this work we address different eigenvalue estimators for static data which are often easier to use in such particular scenarios and discuss benefits and potential limitations. More recently some work also considers eigenvalue estimates in the online or streaming setting (see e.g. [11], for some early work in this line), but here we focus on static data distributions. In particular we address the efficient

T. Villmann et al. (Eds.): WSOM+2024, LNNS 1087, pp. 118–128, 2024.
https://doi.org/10.1007/978-3-031-67159-3_14

estimation of the eigenvalues for large symmetric real valued matrices. Here we consider the psd and non-psd case. The shown techniques can be used to estimate the largest positive and largest negative eigenvalue, the k dominating eigenvalues and eigenvectors, eigenvalue bounds or the psd-ness of a matrix. Eigenvalues and eigenvectors are pivotal in machine learning, enabling key algorithms such as PCA, SVD, spectral clustering, optimization approaches, kernel methods, deep learning and beyond [9,12,15,18]. Experiments provide insights in the behavior of the different methods and can be used as a guidance in practical applications.

2 Eigenvalue Approximation Techniques

In the following we consider a symmetric matrix $\mathbf{A} \in \mathbb{R}^{n \times n}$, where n is the number of samples. A standard eigendecomposition on $\mathbf{A}$ has costs of $O(n^3)$ which is prohibitive for large $n \gg 10000$. As outlined later on this can be reduced by using low rank approximations. The Lanczos iteration, a leading eigendecomposition method, can be applied to low-rank approximated matrices in some frameworks (e.g., Matlab) without fully expanding the matrix.

2.1 Exact Approaches

Many exact methods have been provided to do eigendecomposition, where the modified Lanczos approach is very prominent (see e.g. [2]). As a brief sketch of the idea, the Lanczos iteration aims to minimize the following optimization problem (also known as the Rayleigh quotient):

$$\lambda_{min} = \min_{\mathbf{x} \neq 0} \frac{\langle \mathbf{x}, \mathbf{A}\mathbf{x} \rangle}{\langle \mathbf{x}, \mathbf{x} \rangle},$$

with a symmetric matrix $\mathbf{A} \in \mathbb{R}^{n \times n}$ and its smallest eigenvalue λ_{min} [2]. The actual algorithm does not need access to the explicit matrix, but only to a function that computes the product of the matrix by an arbitrary vector, which is helpful if $\mathbf{A}$ is given as a low-rank approximation, discussed in the second part. The algorithm creates a sequence of orthogonal bases $\{\mathbf{B}_i\}$ formed by i Lanczos vectors $\mathbf{v}_1, \ldots, \mathbf{v}_i$ from *Krylov subspaces* $\mathcal{W}_i(\mathbf{x})$ spanned by vectors $\mathbf{x}, \mathbf{A}\mathbf{x}, \ldots, \mathbf{A}^{i-1}\mathbf{x}$, so that $\mathbf{T}_i = \mathbf{B}_i \mathbf{A} \mathbf{B}_i^T$ has a tridiagonal form, which is computationally easier to handle. It has been shown that the smallest eigenvalue of $\mathbf{T}_i$ is a good approximation to λ_{min} of $\mathbf{A}$ [2]. Therefore, the Lanczos iteration creates the basis $\mathbf{B}_i$ incrementally until convergence and subsequently computes the eigenvalues, further referred to as Ritz values. The convergence rate is heavily related to the overall stiffness of the matrix, but usually, the extreme Ritz values are obtained after $i \approx 2\sqrt{n}$ iterations, leading to good approximations of the corresponding eigenvalues of $\mathbf{A}$ [13]. The implementation of the Lanczos method is numerically challenging, but provides very accurate results at reasonable costs. Another exact alternative is provided by the QR decomposition as shown in Algorithm 1. Both methods are well suited to estimate multiple eigenvectors and corresponding eigenvalues of a matrix, but are also more challenging in the implementation. An alternative is provided by more simple power-iteration based techniques.

Algorithm 1. Eigenvalue Calculation using QR Method

```
function EV_BY_QR(A, maxIterations, tolerance)
    n ← size(A, 1)
    eigenvalues ← zeros(n, 1)
    eigenvectors ← eye(n)
    for k ← 1 to maxIterations do
        [Q, R] ← mgs(A)                          ▷ QR decomposition using MGS
        A ← R · Q                                ▷ Update matrix A
        eigenvectors ← eigenvectors · Q          ▷ Store eigenvectors
        if max(abs(triu(A, 1))) < tolerance then
            break                                ▷ Check for convergence
        end if
    end for
    eigenvalues ← diag(A)                        ▷ Extract eigenvalues from diagonal of A
    return eigenvalues, eigenvectors
end function
```

Algorithm 2. Modified Gram-Schmidt (MGS)

```
function MGS(A)
    [m, n] ← size(A)
    V ← A
    R ← zeros(n, n)
    for i ← 1 to n do
        R(i, i) ← norm(V(:, i))
        V(:, i) ← V(:, i)/R(i, i)
        if (i < n) then
            for j ← i + 1 to n do
                R(i, j) ← V(:, i)' · V(:, j)
                V(:, j) ← V(:, j) − R(i, j) · V(:, i)
            end for
        end if
    end for
    Q ← V      ▷ Orthogonalized matrix Q
    return Q, R
end function
```

2.2 *von Mises*/Power-Iteration

The *von Mises* power iteration method [14] is a widely used technique for approximating the dominant eigenvalue and corresponding eigenvector of a symmetric matrix $\mathbf{A}$. Given a symmetric matrix $\mathbf{A} \in \mathbb{R}^{n \times n}$, the method aims to find the leading eigenpair $(\lambda_1, \mathbf{v}_1)$, where λ_1 is the dominant eigenvalue, and $\mathbf{v}_1$ is the associated eigenvector. The iterative process begins with an initial guess for the eigenvector, denoted as $\mathbf{v}_1^{(0)}$ and is then updated iteratively:

$$\mathbf{v}_1^{(k+1)} = \frac{\mathbf{A}\mathbf{v}_1^{(k)}}{\|\mathbf{A}\mathbf{v}_1^{(k)}\|} \tag{1}$$

Here, $\|\cdot\|$ represents the Euclidean norm. The power iteration is repeated until convergence, typically monitored by the change in the estimated eigenvalue or eigenvector. The dominant eigenvalue is approximated using the Rayleigh quotient:

$$\lambda_1 \approx \frac{\mathbf{v}_1^T \mathbf{A}\mathbf{v}_1}{\mathbf{v}_1^T \mathbf{v}_1} \tag{2}$$

The accuracy of the *von Mises* power iteration method depends on the spectral properties of the matrix $\mathbf{A}$. Convergence is generally faster for matrices with a distinct dominant eigenvalue. Additional eigenvectors and eigenvalues can be calculated by projecting $\mathbf{A}$ into the remaining subspace: $\mathbf{A}' = \mathbf{A} - \langle v_i, v_i \rangle$ and reapplying the *von Mises* iteration. Due to it simplicity the *von Mises* iteration is interesting for systems with limited technical ressources. As we will see in the following it can also be effectively applied to low rank approximated matrices.

2.3 Gershgorin Circles

An alternative method to accelerate the estimations of eigenvalues λ is to approximate the region in which the eigenvalue can be found. The identification of this region can be efficiently achieved by the Gershgorin circle theorem [17]. Let $\mathbf{A} = (a_{ij})$ be a square matrix $(N \times N)$ and $r_i = \sum_{j \neq i} |a_{ij}|$ the row sums of this matrix. Then, within the Gershgorin circle theorem, one may define a disc D_i in the complex plane with center a_{ii} and radius r_i. In [17], it is shown why this can be employed to obtain a valid estimate of the eigenvalues of $\mathbf{A}$. With $D_i = \{z \in \mathbb{C} : |z - a_{ii}| \leq r_i\}$, we obtain ranges that contain the eigenvalues of $\mathbf{A}$: $[a_{ii} - r_i, a_{ii} + r_i]$. Hence one only has to calculate N row-sums and to evaluate the main diagonal of $\mathbf{A}$. The obtained results can be used to find the minimum eigenvalue of $\mathbf{A}$. As an example, consider the following 3×3 matrix for $\mathbf{A}$:

$$\mathbf{A} = \begin{pmatrix} -6 & 1 & -1 \\ 1 & -2 & 5 \\ -1 & 5 & 10 \end{pmatrix} \tag{3}$$

The matrix is symmetric, so all eigenvalues are real. For each row in S, there is one Gershgorin circle defined by its center and its radius:

- D_1 with the center point $c_1 = s_{11} = -6$ and $r_1 = |1| + |-1| = 2$
- D_2 with the center point $c_2 = s_{22} = -2$ and $r_2 = |1| + |5| = 6$
- D_3 with the center point $c_3 = s_{33} = 10$ and $r_3 = |-1| + |5| = 6$

This implicates, all eigenvalues of S must lie in one of the ranges

$$[s_{11} - r_1, s_{11} + r_1] = [-8, -4],\ [s_{22} - r_2, s_{22} + r_2] = [-8, 4],\ [s_{33} - r_3, s_{33} + r_3] = [4, 16].$$

Performing the numerical computation shows that the eigenvalues are approximately $\{-6.6, -3.2, 11.8\}$, all inside the determined ranges. Using the Gershgorin circle approach, we see that the minimum eigenvalue cannot be smaller than the minimum border value, in this example -8, while the right value is ≈ -6.6.

3 Eigenvalue Estimation at Large Scale

In particular for large matrices with $n \gg 10.000$ entries exact eigendecompositions are very costly. As in practical applications also the calculation and storage of a large matrix $\mathbf{A}$ is inconvenient and low rank techniques can be used. Various approaches have been proposed [7] and the Nyström approximation is one of the most prominent low rank techniques. Besides of leading to a simple and low cost approximation of $\mathbf{A}$ by two matrices C and W the Nyström approximation is solely based on landmarks, directly linked to objects from the original $\mathbf{A}$-generating domain. This is often a key requirement in practical use cases.

3.1 Nyström Approximation

The Nyström approximation technique has been proposed in the context of kernel methods in [20] with related proofs and bounds given in [4]. Here, we give a short review of this technique following [6,20]. One well known way to approximate a $N \times N$ Gram matrix, is to use a low-rank approximation. This can be done by computing the eigendecomposition of the kernel $\mathbf{K} = \mathbf{U}\mathbf{\Lambda}\mathbf{U}^\top = \mathbf{A}$, where $\mathbf{U}$ is a matrix, whose columns are orthonormal eigenvectors, and $\mathbf{\Lambda}$ is a diagonal matrix consisting of eigenvalues $\mathbf{\Lambda}_{11} \geq \mathbf{\Lambda}_{22} \geq ... \geq 0$, and keeping only the m eigenspaces which correspond to the m largest eigenvalues of the matrix. The approximation is $\mathbf{K} \approx \mathbf{U}_{N,m}\mathbf{\Lambda}_{m,m}\mathbf{U}_{m,N}$, where the indices refer to the size of the corresponding submatrix. The Nyström method approximates a similarity matrix (originally only for kernels) in a similar way, without computing the eigendecomposition of the whole matrix, which otherwise is an $O(N^3)$ operation.

3.2 Nyström Approximation for Similarities

By the Mercer theorem kernels $k(\mathbf{x}, \mathbf{y})$ can be expanded by orthonormal eigenfunctions ψ_i and non negative eigenvalues λ_i in the form

$$k(\mathbf{x}, \mathbf{y}) = \sum_{i=1}^{\infty} \lambda_i \psi_i(\mathbf{x}) \psi_i(\mathbf{y}).$$

The eigenfunctions and eigenvalues of a kernel are defined as the solution of the integral equation $\int k(\mathbf{y}, \mathbf{x})\psi_i(\mathbf{x})p(\mathbf{x})d\mathbf{x} = \lambda_i \psi_i(\mathbf{y})$, where $p(\mathbf{x})$ is the probability density of $\mathbf{x}$. This integral can be approximated based on the Nyström technique by sampling $\mathbf{x}^k$ i.i.d. according to $p(\mathbf{x})$:

$$\frac{1}{m}\sum_{k=1}^{m} k(\mathbf{y}, \mathbf{x}^k)\psi_i(\mathbf{x}^k) \approx \lambda_i \psi_i(\mathbf{y}).$$

Using this approximation and the matrix eigenproblem equation $\mathbf{K}^{(m)}\mathbf{U}^{(m)} = \mathbf{U}^{(m)}\mathbf{\Lambda}^{(m)}$ of the corresponding $m \times m$ Gram sub-matrix $\mathbf{K}^{(m)}$ we can derive the approximations for the eigenfunctions and eigenvalues of the kernel k

$$\lambda_i \approx \frac{\lambda_i^{(m)}}{m}, \quad \psi_i(\mathbf{y}) \approx \frac{\sqrt{m}}{\lambda_i^{(m)}}\mathbf{k}_y \mathbf{u}_i^{(m)}, \tag{4}$$

where $\mathbf{u}_i^{(m)}$ is the ith column of $\mathbf{U}^{(m)}$. Thus, we can approximate ψ_i at an arbitrary point $\mathbf{y}$ as long as we know the vector $\mathbf{k}_y = (k(\mathbf{x}^1, \mathbf{y}), ..., k(\mathbf{x}^m, \mathbf{y}))^\top$.

For a given $N \times N$ Gram matrix $\mathbf{K}$ we randomly choose m rows and respective columns. The corresponding indices's are also called landmarks, and should be chosen such that the data distribution is sufficiently covered or meta information is included [16]. We denote these rows by $C = \mathbf{K}_{N,m}$. Using the formulas (4) we obtain $\tilde{\mathbf{K}} = \sum_{i=1}^{m} 1/\lambda_i^{(m)} \cdot C\mathbf{u}_i^{(m)}(\mathbf{u}_i^{(m)})^\top C^\top$, where $\lambda_i^{(m)}$ and $\mathbf{u}_i^{(m)}$ correspond to the $m \times m$ eigenproblem. Thus we get, $W = \mathbf{K}_{m,m}^{-1}$ denoting the Moore-Penrose

Algorithm 3. Eigenvalue Decomposition using Nyström Approximation

procedure EIGNYSTRÖM(C, W)
  $Temp \leftarrow W \cdot (C' \cdot C) \cdot W$
  $Temp \leftarrow 0.5 \cdot (Temp + Temp')$
  $[U, A] \leftarrow \text{eig}(Temp)$
  $A(A < \epsilon) \leftarrow 0$
  $B \leftarrow C \cdot (U \cdot \sqrt{A})$
  $[V, A] \leftarrow \text{eig}(B' \cdot B)$
  $A(A < \epsilon) \leftarrow 0$
  $E \leftarrow B \cdot (V \cdot \sqrt{\text{pinv}(A)})$
  $[A, \text{idx}] \leftarrow \text{sort}(\text{diag}((E' \cdot C) \cdot W \cdot (C' \cdot E)), \text{'descend'})$
  $A \leftarrow \text{diag}(A)$
  $E \leftarrow E(:, \text{idx})$
  return E, A
end procedure

pseudoinverse, an approximation of $\mathbf{K}$ as $\tilde{\mathbf{K}} = CWC^\top$. This approximation is exact, if W has the same rank as $\mathbf{K}$.

Using a low rank matrix decomposition the calculation of leading eigenvectors and -values can be substantially accelerated. In some bindings of the Lanzcos iteration low rank matrices are supported as well and the full matrix expansion can be avoided. As shown in [6,10] the Nyström approximation can also be applied to non-psd similarity functions. The accuracy of such a low rank eigendecomposition shown in Algorithm 3 (original derived in [6]) depends in general less on the approach itself, but on the accuracy of the matrix approximation and the characteristics of the eigenspectrum of the original matrix A. See also Table 2 where the results for the low-rank Lanczos method (EV-Ny-LZ) are identical (upto numerical precision) to the Nyström based *von Mises* approximation. A more detailed analysis of how the landmark selection strategy affects the accuracy of approximating both the eigenvalues and eigenvectors of the matrix is provided in [6,16,20].

3.3 *von Mises* Iteration using Nyström

Power iteration methods are easy to implement and often sufficiently accurate. The *von Mises* iteration can be easily modified to be applied on a low rank approximated matrix by using standard concepts from linear algebra (see Algorithm 5). If only a few eigenvectors and -values are to be estimated the *von Mises* iteration is often good enough. With some additional observations it can also be used to approximate the largest positive and the largest negative eigenvalue with a few simple steps as shown in Algorithm 4. Let λ_m be the largest (positive) eigenvalue of a matrix A approximated e.g. by *von Mises* (EV-M). We can than obtain the largest negative eigenvalue of A by inverting A and applying a shift correction: $\lambda_t = \text{VonMises}(-(A - I\lambda_m))$. The largest negative eigenvalue is finally obtained as $\lambda_n = -\lambda_t + \lambda_m$.

Algorithm 4. Largest Eigenvalues using *von Mises* Iteration

```
function LargestEigenvaluesNy(C, W)
    [n, m] ← size(C)
    [e, eMax] ← VonMisesNy(C, W)        ▷ Largest positive eigenvector/-value
    WI ← −W                             ▷ Get the largest negative eigenvalue
    x ← rand(n, 1)
    xOld ← x + 1
    k ← 0
    while Σ |xOld − x| > 1e − 8 and k < 1000 do
        xOld ← x
        x ← C · ((C′ · x)′ · WI)′ + x · eMax
        x ← x/norm(x)
        k ← k + 1
    end while                           ▷ Calculate eigenvalue using the eigen equation
    eT ← x\(C · ((C′ · x)′ · WI)′ + x · eMax)
    eMin ← −(eT − eMax)
    return eMax, eMin
end function
```

3.4 Estimating the Gershgorin Circle Bounds at Large Scale

The Gershgorin Theorem needs the diagonal elements and the row sums of a given matrix A. Using matrix low rank approximations like Nyström this can be done with linear costs as follows:

1. For the k-th row sum of the matrix K, denoted as $\sum_i [\hat{A}]_{k,i}$:

$$\sum_i [\hat{A}]_{k,i} = \left(\sum_i CW^{-1}\right) [C']_{\cdot,k}$$

2. For the diagonal elements of the matrix A, denoted as $\mathrm{diag}(A)$:

$$\mathrm{diag}(A) = \sum \left((W^{-1}C') \odot C'\right)$$

The respective values are entered into the Gershgorin circles to obtain the final bounds. For the shifted Gershgorin circle the modification is not so straight forward but can be done using similar techniques. An overview of the discussed methods is given in Table 1.

Table 1. Comparison of Eigenvalue Approximation Methods

Method	Complexity	Pros	Cons
Lanczos	$O(n^2)$	High accurate	Numerically challenging
von Mises	$O(n^2)$	Simple to implement	Slow convergence for smooth spectra
Gershgorin Circle	$O(n)$	Fast, easy to compute	Only bounds, less precise
Nyström Approximation	$O(m^2 n)$	Efficient for large matrices	Crucial landmark selection
QR Method	$O(n^3)$	Very accurate	Computationally intensive

Algorithm 5. *von Mises* Iteration using Nyström Approximation

procedure VONMISESNY(C, W)
  $n \leftarrow \text{size}(C, 1)$
  $x1 \leftarrow \text{rand}(n, 1)$
  $xOld \leftarrow x1 + 1$
  $k \leftarrow 0$
  while $\sum |xOld - x1| > 1e - 8$ **and** $k < 1000$ **do**
    $xOld \leftarrow x1$
    $x1 \leftarrow C\left(((C' \cdot x1)' \cdot W)'\right)$
    $x1 \leftarrow \frac{x1}{\|x1\|}$
    $k \leftarrow k + 1$
  end while
  $eigenvalue \leftarrow x1 \setminus C\left(((C' \cdot x1)' \cdot W)'\right)$
  return $x1, eigenvalue$
end procedure

4 Experiments

We analyze the considered eigenvalue approximation methods for different datasets with very different characteristics.

1. **Checkerboard** (DS-1) is an artificial data set of 900 2d point in overlapping point clouds. The indefinite similarity matrix was obtained by $\text{ker}(\mathbf{x}, \mathbf{y}) = \tanh(0.1 + 0.5 \cdot \langle \mathbf{x}, \mathbf{y} \rangle^2)$. By modifying the parameters of tanh the indefiniteness of the similarity matrix can be modified rather arbitrary. Using this particular parametrization the rank of the matrix is ≈ 200.
2. The **Protein** dataset (DS-2) comprises 213 protein sequences and their associated indefinite sequence-alignment similarities, enabling protein sequence comparison and classification (see e.g. [16]).
3. **Chromosomes:** The Copenhagen chromosomes dataset (DS-3), a cytogenetics classification benchmark with 4,200 grey-valued images, transforms chromosomal silhouettes into strings. These strings are compared using edit distance (insertion/deletion costs 4.5) to derive indefinite similarities [16].

The shown results are typical examples observed for a wide set of sample data. In particular one can observe that eigenvalues approximated on an exact matrix are in general identical (upto numerical precision) regardless the used method (Lanzcos or Power Iteration). The precision is however often controlled by the number of iterations e.g. in the *von Mises* iteration. If the eigengap (the distance between two consecutive eigenvalues) is large the power iteration method show fast convergence. If the eigenspectrum is smooth the convergence can be very slow. The accuracy for the Nyström approximated matrices depends very much on the precision of the matrix approximation and not on the particular method (which showed only very small variations). There is much work on addressing the parametrization of low rank methods (see e.g. [7]) and we do not detail such a study here. Some more recent work on analyzing numerical errors in eigenvalue estimations can be found in [5]. For simplicity the number of landmarks is just set as $m = \sqrt{N}$ and randomly chosen. Please note that approximating a matrix could make it psd. In Table 2 we show results for approximating the largest positive and the largest negative eigenvalues using exact and approximated techniques. If we only require a rough estimate of the eigenvalues we can also use the Gershgorin circle theorem (columns EV-G,EV-SG). One can see that in some cases (for DS1 and DS2) the estimation using Gershgorin is at least roughly in a valid range. The Gershgorin bounds are not tight. As recently shown in [3] this can be improved by the shifted Gershgorin circle (EV-SG) under some moderate assumptions leading to eigenvalue estimates which are better, but still not very accurate[1] Depending on the scenario it may however be sufficient to consider the bounds only. This may for example be a cheap test whether the matrix is psd, where an exact solution to test for positive definiteness is costly and could be obtained using the Cholesky decomposition. While the results in Table 2 are only done for moderate N, similar observations can be found also for very large N. Using techniques like Nyström the costs of the eigenvalue calculations scale only linear in the number of points (but cubic in the number of landmarks). In practical use cases the meaningful rank of a problem is often only between a few to multiple hundred landmarks. Accordingly eigenvalue decompositions of matrices upto $k, k > 0$ million samples is possible already on standard equipment. Initial tests to estimate the largest eigenvalue for DS-1 with 4.5 million samples and 100 landmarks can be done using Lanczos in less than a minute and with *von Mises* in 20 min. While Lanczos is in general linked to more complicated, potentially expensive math libraries, the *von Mises* iteration is simple to implement and often sufficient.

Table 2. Exemplary differences in estimating the dominating eigenvalue.

DS	EV-Exact	EV-M	EV-Ny-LZ	EV-Ny-M	EV-G	EV-SG
DS-1	−93.59	−93.59	−262.57	−262.57	−802.12	−358.16
DS-2	−12.43	−12.43	−12.83	−12.83	−139.09	−78.60
DS-3	−4.41	−4.24	$-10e-14$	$10e-6$	−2243.40	−1073.25
DS-1	648.19	648.19	576.7050	576.7050	804.12	–
DS-2	72.00	72.00	65.94	65.94	141.0858	–
DS-3	1983.09	1983.09	1865.97	1865.97	2245.40	–

5 Conclusions

The paper investigated various methods to approximate the dominant eigenvalues of matrices, essential in numerous scientific and engineering applications. The

[1] Maximum bound for shifted Gershgorin skipped as these are known to be inaccurate.

explored techniques contained the *von Mises* iteration, the modified Lanczos iteration as an effective method for symmetric matrices and two versions of the Gershgorin circle theorem to get rough estimates of eigenvalue bounds. To address large scale matrices we discussed or provided extensions of the methods using the Nyström approximation. These diverse methods contribute to the toolkit for solving large-scale eigenvalue problems, each tailored to different matrix characteristics and specific domain or computational requirements.

References

1. Anderson, E., et al.: LAPACK Users' Guide, 3rd edn. SIAM, Philadelphia (1999)
2. Cullum, J., Willoughby, R.: Lanczos Algorithms for Large Symmetric Eigenvalue Computations, vol. 1. SIAM, Philadelphia (PA) (2002)
3. DeVille, L.: Optimizing gershgorin for symmetric matrices. Linear Algebra Appl. **577**, 360–383 (2019)
4. Drineas, P., Mahoney, M.W.: On the nyström method for approximating a gram matrix for improved kernel-based learning. J. Mach. L. Res. **6**, 2153–2175 (2005)
5. Gedicke, J.M.: On the Numerical Analysis of Eigenvalue Problems. Thesis, Humboldt-Univ. Berlin, Berlin (2013). https://tinyurl.com/yehhepnt
6. Gisbrecht, A., Schleif, F.: Metric and non-metric proximity transformations at linear costs. Neurocomputing **167**, 643–657 (2015)
7. Kumar, K., Schneider, J.: Literature survey on low rank approximation of matrices. Linear Multilinear Algebra **65**(11), 2212–2244 (2017)
8. Münch, M., Raab, C., Schleif, F.: Encoding of indefinite proximity data: a structure preserving perspective. In: Marsico, M.D., di Baja, G.S., Fred, A.L.N. (eds.) ICPRAM 2020. LNCS, vol. 12594, pp. 112–137. Springer, Cham (2020). https://doi.org/10.1007/978-3-030-66125-0_7
9. Münch, M., Röder, M., Heilig, S., Raab, C., Schleif, F.: Static and adaptive subspace information fusion for indefinite heterogeneous proximity data. Neurocomputing **555**, 126635 (2023)
10. Oglic, D., Gärtner, T.: Scalable learning in reproducing kernel Krein spaces. In: International Conference on Machine Learning, pp. 4912–4921. PMLR (2019)
11. Oja, E., Karhunen, J.: On stochastic approx of the eigenvectors and -values of the expectation of a random matrix. J. Math. Anal. and Appl. **106**(1), 69–84 (1985)
12. Olteanu, M., Villa-Vialaneix, N., Cierco-Ayrolles, C.: Multiple kernel self-organizing maps. In: 21st European Symposium on Artificial Neural Networks, ESANN 2013, Bruges, Belgium, April 24–26, 2013 (2013)
13. Parlett, B.N., Simon, H., Stringer, L.: On estimating the largest eigenvalue with the Lanczos algorithm. Math. Comput. **38**(157), 153–165 (1982)
14. Rabbani, T., Jain, A., Rajkumar, A., Huang, F.: Practical and fast momentum-based power methods. In: Bruna, J., Hesthaven, J.S., Zdeborová, L. (eds.) Mathematical and Scientific Machine Learning. PMLR, vol. 145, pp. 721–756. PMLR (2021)
15. Schleif, F.-M., Villmann, T., Hammer, B., Schneider, P., Biehl, M.: Generalized derivative based kernelized learning vector quantization. In: Fyfe, C., Tino, P., Charles, D., Garcia-Osorio, C., Yin, H. (eds.) IDEAL 2010. LNCS, vol. 6283, pp. 21–28. Springer, Heidelberg (2010). https://doi.org/10.1007/978-3-642-15381-5_3
16. Schleif, F., Gisbrecht, A., Tiňo, P.: Supervised low rank indefinite kernel approximation using minimum enclosing balls. Neurocomputing **318**, 213–226 (2018)

17. Varga, R.S.: Geršgorin and His Circles. Springer Series in Computational Mathematics, vol. 36. Springer, Heidelberg (2004)
18. Villmann, T., Haase, S., Kaden, M.: Kernelized vector quantization in gradient-descent learning. Neurocomputing **147**, 83–95 (2015)
19. Wang, M., Fu, W., He, X., Hao, S., Wu, X.: A survey on large-scale machine learning. IEEE TKDE **34**, 2574–2594 (2020)
20. Williams, C., Seeger, M.: Using the Nyström method to speed up kernel machines. In: Advance in NIPS, pp. 682–688. MIT Press (2001)

Enhancing LDA Method by the Use of Feature Maximization

Jean-Charles Lamirel(✉)

Université de Strasbourg, SYNALP-LORIA, 615 Rue du Jardin-Botanique, 54506 Vandœuvre-lès-Nancy, France
lamirel@loria.fr

Abstract. Topic modeling is a key technique for understanding the content of collections of scientific papers. However, commonly used methods like LDA (Latent Dirichlet Allocation) have significant drawbacks, including complex parameter settings. Additionally, these methods often yield low-quality results. Therefore, improving the outcomes of topic modeling is a crucial goal. In this paper, we compare the performance of LDA with a recent topic modeling approach we have developed, which relies on a combination of neural clustering and feature maximization (abbreviated as CFMf). Subsequently, we demonstrate how the feature ranking component of the CFMf method can be used to substantially enhance the performance of LDA, regardless of the number of topics. We also highlight the benefits of post-processing the clustering results before modeling topics in the CFMf approach. Our reference dataset consists of 16,917 full-text articles on the philosophy of science.

Keywords: Topic model · CFMf · LDA · Top2Vec · Clustering · Topic coherence

1 Introduction

Computationally mining the unstructured textual content of scientific text is an effective way to analyze large corpora that are too extensive for close reading. In this context, topic modeling can be used to reliably infer the thematic content of publications, enabling the identification of dominant research topics within specific scientific disciplines and their evolution over time [1]. One well-established approach to topic modeling is the Latent Dirichlet Allocation (LDA) model [2], a statistical bag-of-words machine learning model, along with its various adaptations. Recent works have successfully developed alternatives to statistical generative models like LDA by using a combination of clustering with feature selection for topic modeling. For example, the CFMf approach proposes a joint use of neural clustering and a feature maximization metric (in short, Fmax [4]) for that purpose [13]. Alternatively, the emergence of word embeddings and large language models has introduced a fundamentally different approach to topic extraction, exemplified by techniques like Top2Vec, which is based on a joint use of word and document embeddings [3].

T. Villmann et al. (Eds.): WSOM+2024, LNNS 1087, pp. 129–138, 2024.
https://doi.org/10.1007/978-3-031-67159-3_15

On the one hand, LDA is known to be dependent on hyperparameters, which must be adjusted for each new experiment. Additionally, LDA relies on the questionable assumption that extracted topics must be mutually independent, a condition that may not hold true for complex data. On the other hand, Top2Vec has a significant drawback in that it relies on a language model that must be pre-trained, introducing dependence on the specific language being processed. One of the main advantages of the CFMf method is that it is parameter-free and language-agnostic method that does not suffer from any of the aforementioned drawbacks.

We recently compared the LDA method with CFMf and Top2Vec methods using four different performance measures[1] and a large range of topic numbers [13]. Figures 1 and 2 summarizes the values of the different performance measures for the experimented methods. The average ranking of the methods regarding their performance for the different measures (see Table 1) confirmed that the CFMf method was the best overall performer, making it a very promising alternative to LDA and Top2Vec. This also indicates that statistical models can still prove useful despite the rise in popularity of embeddings and large language models.

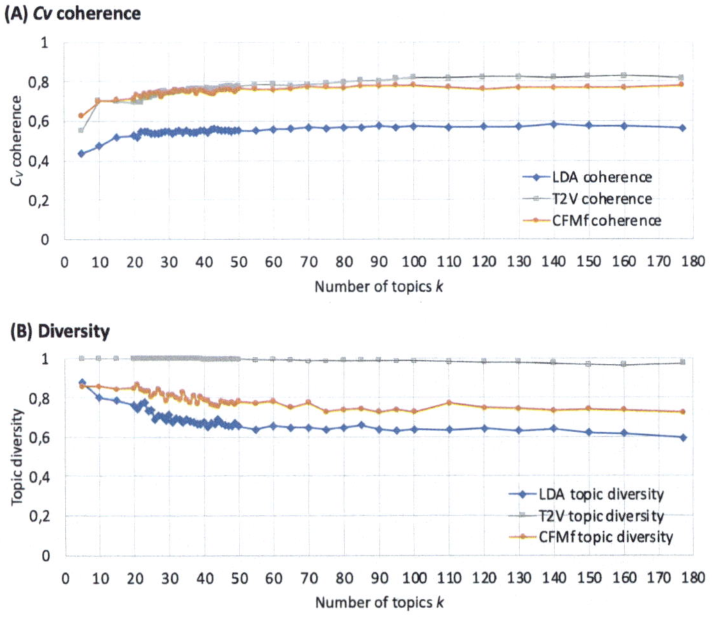

Fig. 1. Results of the C_v coherence (A) and the Topic diversity (B) measures for the different topic models over a large range of topic numbers.

[1] See reference [13] for a more precise description of the topic diversity, inner recall, and joint inner recall measures.

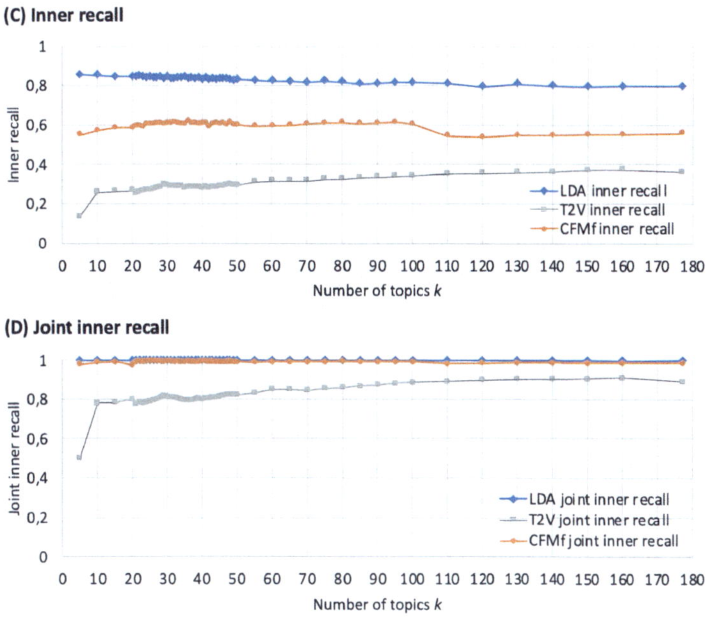

Fig. 2. Results of the Inner Recall (C) and Joint Inner Recall (D) measures for the different topic models over a large range of topic numbers.

Table 1. Average rank of the methods for the 4 performance measures.

	Coherence	Diversity	Inner Recall	Joint Inner Recall	Average Rank
LDA	3	2	2	1	1.75
Top2Vec	1	2	3	3	2.25
CFMf	1	1	2	1	**1.25**

Our previous findings [5,13] also highlighted another significant weakness of LDA: it tends to generate topics with lower coherence compared to other methods, which substantially impacts interpretability. Therefore, in this new paper, we propose a novel strategy that involves utilizing the Fmax metric to improve the results of LDA topic modeling.

Some clustering issues with high-dimensional data have also been identified in our previous experiments [13], notably the existence of incoherent clusters in the generated clustering models due to the processing of high-dimensional data. Therefore, in this paper, we also emphasize the importance of post-processing the clustering results before modeling topics in the CFMf approach.

Next, we introduce the corpus, outline the two topic models, and detail the series of comparative tests we performed. Subsequently, we present our findings. Finally, we discuss potential directions for further research.

2 Corpus and Methods

Our experimental corpus has been used in many of our previous experiments [6, 13]. It comprises the full-text research articles from eight prominent philosophy of science journals. The corpus covers a large period from 1930 to 2017 and includes a total of 16,917 documents. The corpus preprocessing steps are exhaustively described in [13]. These steps resulted in producing a document-term matrix (DTM) of word frequencies with dimensions 16,917 × 23,762.

In our new experiment, two methods are considered, namely the CFMf method and the standard LDA method.

The CFMf method performs an additional preprocessing step of (log of tf)-idf weighting to the document-term matrix in order to reduce the influence of the most widespread words within the corpus and to jointly cope with the known imbalanced frequencies of words inside documents, according to the Zipf law.

A crisp clustering process using Growing Neural Gas (GNG) neural clustering is then applied to the weighted matrix. This process is followed by a word selection and word ranking process in each cluster in order to suitably extract cluster-related topics. This latter process relies on data associated with clusters and is based on the feature F1-measure. More precisely, let us consider a partition C which results from a clustering method applied to a dataset D represented by a group of features F, the feature recall $FR_c(f)$ and the feature predominance $FP_c(f)$ and the feature F1-measure $FF1_c(f)$, are defined as follows:

$$FR_c(f) = \frac{\sum_{d \in c} W_d^f}{\sum_{c \in C} \sum_{d \in c} W_d^f} \tag{1}$$

$$FR_c(f) = \frac{\sum_{d \in c} W_d^f}{\sum_{f' \in F_c, d \in c} W_d^{f'}} \tag{2}$$

with

$$FF1_c(f) = \frac{2((FR_c(f) \times FP_c(f))}{(FR_c(f) + FP_c(f)))} \tag{3}$$

where W_d^f represents the weight of the feature f for the data d and F_c represents all the features present in the data associated to the cluster c.

Word selection in clusters is an additional step that is performed by suppressing words with F1-measures below the global average of F1-measures across all clusters. Such words are considered to be non-discriminative for any cluster.

LDA (Latent Dirichlet Allocation) is a generative statistical model that considers documents, as well as the embedding dataset, as mixtures of topics. The method assumes that the underlying topic structure is hidden and constrained by sparse Dirichlet priors representing external hyperparameters. After setting these hyperparameters, the topic structure is revealed through an iterative process, such as Gibbs sampling, starting by assigning random probability distributions to words within topics and topics within documents.

In contrast to the CFMf approach, LDA modeling is applied directly to the Document-Term Matrix (DTM) of word frequencies (with dimensions of $16{,}917 \times 23{,}762$). It does not rely on any additional weighting.

For conducting our overall experiment we process in the following way, as presented in Fig. 3:

1. The optimal number of topics is determined using the CFMf method in conjunction with a strategy for detecting the optimal clustering model based on contrast derived from the Fmax metric proposed in [14]. For that purpose we generate clustering models in a large range of value of K ($K = 2\,to\,200$) and apply the contrast measures on all the generated models. This process results in an optimal model with $K = 47$ topics. Incoherent topics, defined as those with a top-1 word f1-measure lower than the average f1-measure of the top-k words computed across all topics in the model, which the value of k experimentally set to 5 as described in Algorithm 1, are identified and removed. The data associated with these supposedly highly incoherent topics are then reallocated to the remaining topics. The resulting topic model, with $K = 34\,topic$, is considered the reference size for further experiments. This model is named *GNG-FMAX-47C34*.
2. A CFMf model with 34 topics, for which no topic reorganisation is performed, is computed. This model is named *GNG-FMAX-34*.
3. A standard LDA model with $K = 34$ topic is computed as well. This model is named *LDA-34*.
4. The LDA topic model is crisp clustered by assigning data to their most dominant topic. The resulting cluster model is then processed using the Fmax metric to extract the top words of the cluster, thereby reassigning the top words to the original LDA model. This model is named *LDA-FMAX-34*.

The resulting topic models are evaluated using coherence performance metrics, specifically focusing on the top-k words of each topic while varying the value of k. The C_v coherence measure, as defined in [7] and implemented in the widely-used Gensim package in Python[2], is employed for this purpose. C_v assesses the co-occurrence patterns of a specified number of top words (typically between 10 and 20) within a sliding window of fixed size (typically 110). These co-occurrence patterns are then utilized to compute the Normalized Pointwise Mutual Information (NPMI) [12] among the top words, generating vectors for each. The coherence score is derived by calculating the arithmetic mean of the cosine similarities between each top word vector and the aggregate of all top word vectors. Topic coherence gauges how semantically related the top W words are within each topic in a topic model (or cluster, as in methods like CFMf). Higher coherence values indicate that the top words within each topic are more semantically cohesive. The C_v coherence measure is among the most commonly used metrics for this type of evaluation.

[2] https://pypi.org/project/gensim/.

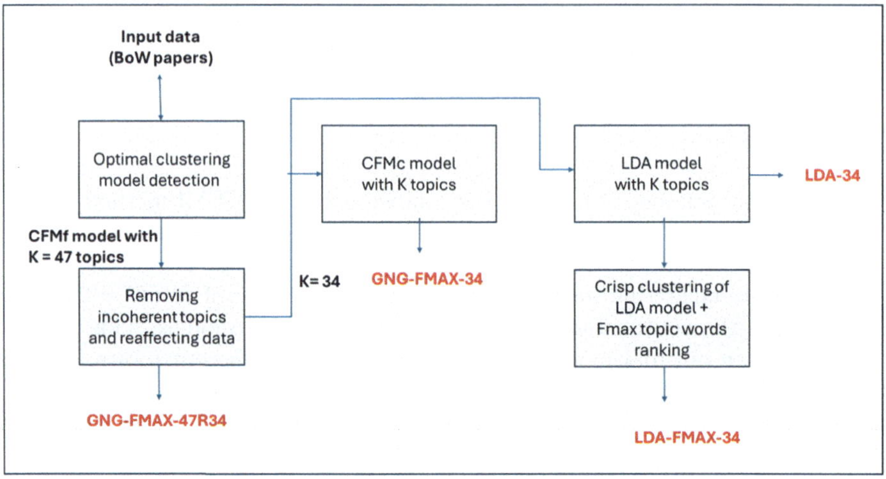

Fig. 3. Overview of the experimental process.

$RF_1(c, i)$ provides the F1-measure of the top word of rank i in the cluster c.
$Aff(c, d)$ a function that associates the data d to the cluster c.
$Dist(v_1, v_2)$ a function that calculates the distance between vectors v_1 and v_2.
$V(c)$ a function that provides the profile vector of cluster c.
d_c the set of data associated to cluster c.
Data: C a clustering model with n clusters
Data: k the number of top words used for F1-average ($k = 5$)
Result: R a clustering model with $p \leq n$ clusters
/* Calculating the F1-average of the top-k words for the model */
for $c \in C$ **do**
  for $i = 1$ *to* k **do**
    $AvF_1 \leftarrow AvF_1 + RF_1(c, i)$
  end
  $AvF_1 \leftarrow \frac{AvF_1}{k \times \|C\|}$
end
/* Suppressing incoherent topics */
$R = \emptyset$ **for** $c \in C$ **do**
  if $RF_1(c, 1) > AvF_1$ **then**
    $R = R \cup c$
  end
end
/* Reclassing data of suppressed topics in remaining topics */
for $c \in C$ **do**
  for $d \in d_c$ **do**
    $w = argmin_{r \in R} Dist(V(r), d)$
    $Aff(w, d)$
  end
end
Algorithm 1: Suppressing incoherent topics and reassociating their data.

3 Results

Results in terms of topic coherence of the *GNG-FMAX-47R34*, *GNG-FMAX-34*, *LDA-34* and *LDA-FMAX-34* models are graphically displayed on Fig. 4, The lowest coherence is obtained when LDA is used in its standard mode, utilizing word probabilistic ranking in topics (*LDA-34 model*). The highest coherence is achieved with the CFMf model when the optimal clustering model is post-processed to eliminate incoherent topics (*GNG-FMAX-47C34 model*). The obtained coherence of more than 0.8 is extremely high compared to the usual values reported in the literature. Moreover, the coherence increase compared to the standard LDA model is as high as 53%. The optimally CFMf post-processed model, *GNG-FMAX-47C34*, is slightly better (by 1%) than an equivalent CFMf model in terms of the number of topics, *GNG-FMAX-34*, these latter being obtained directly without post-processing. This demonstrates the benefit of post-processing the clustering results by eliminating incoherent clusters and reallocating data to the remaining clusters. The enhanced LDA model using Fmax ranking (*LDA-FMAX-34* model) proves to be significantly better in terms of coherence than the standard LDA model, *LDA-34*. The improvement is substantial, i.e. around 43%. This experiment thus shows that word ranking based on the probabilistic approach provided by LDA is far from optimal. However, the enhanced LDA model still has slightly lower coherence (6–7% coherence loss) compared to the models produced by the CFMf method. This illustrates the interest of exploiting clustering instead of using a generative approach for topic modeling.

The Fig. 5 shows that the strategy of starting with a larger optimal model, removing incoherent topics using the F1-measure, and reclassifying data into the remaining topics (*GNG-FMAX-47R34 model*) with Algorithm 1, whilst increasing coherence (see Fig. 4), has the additional benefit of rebalancing class sizes and eliminating incoherent topics that gather weakly described documents.

Last but not least, as illustrated in Fig. 4, all the models using Fmax ranking, specifically *GNG-FMAX-47C34*, *GNG-FMAX-34*, and *LDA-FMAX-34*, tend to exhibit a lower coherence decrease as the number k of top-k words considered in topic descriptions increases. This proves that the consistency of Fmax ranking is higher than that of probabilistic ranking.

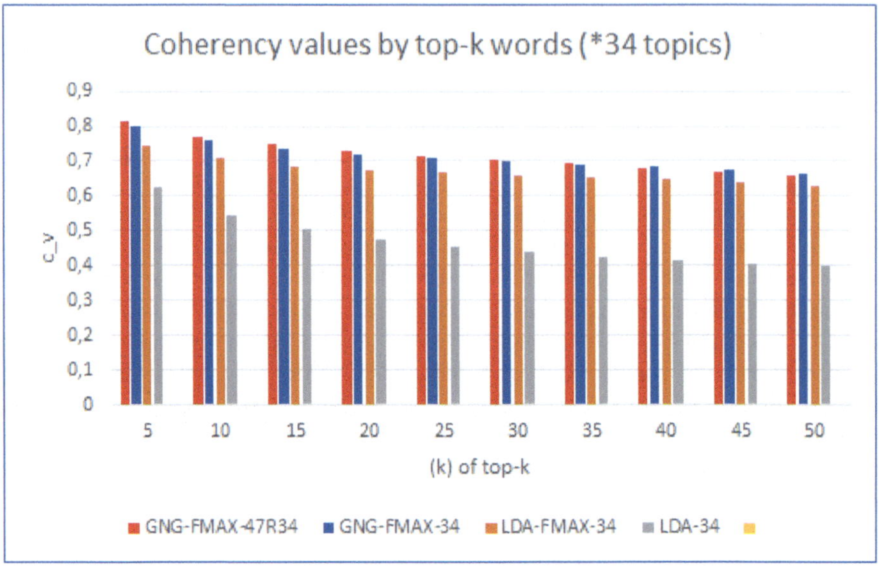

Fig. 4. Comparison histograms of C_v coherence between topic models. Results of the four performance measures implemented on each topic model (*GNG-FMAX-47C34*, *GNG-FMAX-34*, *LDA-34* and *LDA-FMAX-34*) as a function of the number k of top-words.

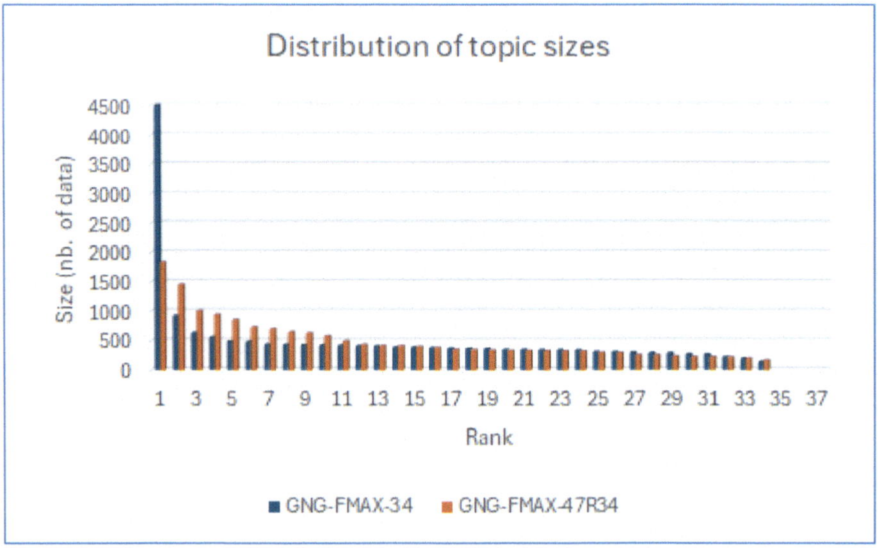

Fig. 5. Distribution of topic sizes for 34 topics models with an without suppressing topics from a larger model and reclassing their associated data.

4 Conclusion

Based on observations from our previous experiments, we have presented a promising new approach for improving an existing and extensively used state-of-the-art topic modeling method, namely LDA, by employing a word ranking strategy utilized in another topic model we recently developed. Our initial results indicate that a significant increase in model coherence can be achieved through this approach, consequently enhancing the interpretability of the LDA models. This serves as initial evidence that there is potential for enhancing the operating mode of well-known methods. Our supplementary experiments also demonstrate that, in the case of the CFMf method, clustering results can be optimized and interpretability enhanced by eliminating incoherent clusters through appropriate post-processing of the clustering model. However, further investigation is necessary to reliably estimate clustering incoherencies in high-dimensional clustering models without fixing parameters. In the near future, we plan to apply Fmax and GNG clustering in the context of methods based on embeddings, such as Top2Vec. This is expected to improve the interpretability of the topic models generated by such methods and address inconsistencies in results caused by rare topics appearing in the generated models, as observed in our previous experiments. Furthermore, we intend to conduct additional qualitative analysis of the results produced by these methods and expand the scope of our experiments to include other datasets. We will also explore the impact of lexicon reduction on the three types of models. Lastly, in refining the operational mode of an improved CFMf model, we will investigate specific strategies to combine F1-measure and contrast ranking to further optimize topic descriptions.

References

1. Griffiths, T.L., Steyvers, M.: Finding scientific topics. Proc. Natl. Acad. Sci. **101**(suppl 1), 5228–5235 (2004)
2. Blei, D.M., Ng, A.Y., Jordan, M.I.: Latent Dirichlet allocation. J. Mach. Learn. Res. **3**, 993–1022 (2003)
3. Angelov, D.: Top2Vec: distributed representations of topics (2020). http://arxiv.org/abs/2008.09470
4. Lamirel, J.-C., Cuxac, P., Chivukula, A.S., Hajlaoui, K.: Optimizing text classification through efficient feature selection based on quality metric. J. Intell. Inf. Syst. **45**(3), 379–396 (2015)
5. Lamirel, J.-C., Chen, Y., Cuxac, P., Al Shehabi, S., Dugué, N., Liu, Z.: An overview of the history of Science of Science in China based on the use of bibliographic and citation data: a new method of analysis based on clustering with feature maximization and contrast graphs. Scientometrics **125**(3), 2971–2999 (2020)
6. Malaterre, C., Lareau, F.: The early days of contemporary philosophy of science: novel insights from machine translation and topic-modeling of non-parallel multilingual corpora. Synthèse **200**(3), 242 (2022)
7. Röder, M., Both, A., Hinneburg, A.: Exploring the space of topic coherence measures. In: Proceedings of the Eighth ACM International Conference on Web Search and Data Mining - WSDM 2015, pp. 399–408 (2015)

8. Mikolov, T., Sutskever, I., Chen, K., Corrado, G.S., Dean, J.: Distributed representations of words and phrases and their compositionality. In: Advances in Neural Information Processing Systems, vol. 26 (2013)
9. Fritzke, B.: A growing neural gas network learns topologies. In: Advances in Neural Information Processing Systems, vol. 7 (1994)
10. McInnes, L., Healy, J., Melville, J.: UMAP: uniform manifold approximation and projection for dimension reduction (2020). http://arxiv.org/abs/1802.03426
11. Campello, R.J.G.B., Moulavi, D., Sander, J.: Density-based clustering based on hierarchical density estimates. In: Pei, J., Tseng, V.S., Cao, L., Motoda, H., Xu, G. (eds.) PAKDD 2013. LNCS (LNAI), vol. 7819, pp. 160–172. Springer, Heidelberg (2013). https://doi.org/10.1007/978-3-642-37456-2_14
12. Gerlof, B.: Normalized (pointwise) mutual information in collocation extraction. In: Proceedings of GSCL, vol. 30, pp. 31–40 (2009)
13. Lamirel, J.-C., Malaterre, C., Lareau, F.: CFMf topic-model: comparison with LDA and Top2Vec. Scientometrics (2024)
14. Dugué, N., Lamirel, J.-C.: Evaluating clustering quality using features salience: a promising approach. Neural Comput. Appl. **19**, 12939–12956 (2021)

Explaining Neural Networks - Deep and Shallow

Barbara Hammer(✉)

Machine Learning Group, CITEC Bielefeld University, Bielefeld, Germany
bhammer@techfak.uni-bielefeld.de

Abstract. Variable importance determination refers to the challenge to identify the most relevant input dimensions or features for a given learning task and quantify their relevance, either with respect to a local decision or a global model. Feature relevance determination constitutes a foundation for feature selection, and it enables an intuitive insight into the rational of model decisions. Indeed, it constitutes one of the oldest and most prominent explanation technologies for machine learning models with relevance for both, deep and shallow networks. A huge number of measures have been proposed such as mutual information, permutation feature importance, deep lift, LIME, GMLVQ, or Shapley values, to name just a few.

Within the talk, I will address recent extensions of feature relevance determination, which occur as machine learning models are increasingly used in everyday life. Here, models face an open environment, possibly changing dynamics, and the necessity of model adaptation to account for changes of the underlying distribution. At present, feature relevance determination almost solely focusses on static scenarios and batch training. In the talk, I will target the question of how to efficiently and effectively accompany a model which learns incrementally by feature relevance determination methods [1,3]. As a second challenge, features are often not mutually independent, and the relevance of groups rather than single features should be judged. While mathematical models such as Shapley values take feature correlations into account for individual additive feature relevance terms, it is unclear how to efficiently and effectively extend those to groups of features. In the talk, I will discuss novel methods for the efficient computation of feature interaction indices [2,4].

References

1. Fumagalli, F., Muschalik, M., Hüllermeier, E., Hammer, B.: Incremental permutation feature importance (iPFI): towards online explanations on data streams. Mach. Learn. **112**(12), 4863–4903 (2023)
2. Fumagalli, F., Muschalik, M., Kolpaczki, P., Hüllermeier, E., Hammer, B.: SHAP-IQ: unified approximation of any-order shapley interactions. In: NeurIPS (2023)

T. Villmann et al. (Eds.): WSOM+2024, LNNS 1087, pp. 139–140, 2024.
https://doi.org/10.1007/978-3-031-67159-3_16

3. Muschalik, M., Fumagalli, F., Hammer, B., Hüllermeier, E.: iSAGE: an incremental version of SAGE for online explanation on data streams. In: Koutra, D., Plant, C., Gomez Rodriguez, M., Baralis, E., Bonchi, F. (eds.) ECML PKDD 2023. LNCS, vol. 14171, pp. 428–445. Springer, Cham (2023). https://doi.org/10.1007/978-3-031-43418-1_26
4. Muschalik, M., Fumagalli, F., Hammer, B., Hüllermeier, E.: Beyond treeshap: efficient computation of any-order shapley interactions for tree ensembles. In: AAAI (2024)

FairGLVQ: Fairness in Partition-Based Classification

Felix Störck(✉), Fabian Hinder, Johannes Brinkrolf, Benjamin Paassen, Valerie Vaquet, and Barbara Hammer

Bielefeld University - Cognitive Interaction Technology (CITEC), Inspiration 1, 33619 Bielefeld, Germany
{fstoerck,fhinder,jbrinkro,bpaassen,vvaquet, bhammer}@techfak.uni-bielefeld.de

Abstract. Fairness is an important objective throughout society. From the distribution of limited goods such as education, over hiring and payment, to taxes, legislation, and jurisprudence. Due to the increasing importance of machine learning approaches in all areas of daily life including those related to health, security, and equity, an increasing amount of research focuses on fair machine learning. In this work, we focus on the fairness of partition- and prototype-based models. The contribution of this work is twofold: 1) we develop a general framework for fair machine learning of partition-based models that does not depend on a specific fairness definition, and 2) we derive a fair version of learning vector quantization (LVQ) as a specific instantiation. We compare the resulting algorithm against other algorithms from the literature on theoretical and real-world data showing its practical relevance.

Keywords: Fair Machine Learning · Fairness · Partition-based Models · Prototype-based Models · Fair LVQ

1 Introduction

Whenever the outputs of classification models impact human lives, fairness is a crucial concern. For example, if models suggest grades in school, score applications in human resources, or distribute scarce resources like in water networks, individuals and groups have good reason to demand equal treatment [10,16,17]. However, fairness alone is insufficient for the requirements of modern socio-technical systems. Especially in the European context, models also need to facilitate explainability and transparency to be eligible for the label "trustworthy artificial intelligence" as introduced by the EU [3] and as codified in the European AI act [7]. Further requirements may be technical in nature, such as compatibility with streaming data or data sparsity.

While extensive prior work on fairness in machine learning exists, most of it has focused solely on the relation of fairness and accuracy [10,17], whereas other requirements have been mostly ignored. Prior efforts to make interpretable models more fair have been almost exclusively focused on decision trees [4,8,12].

T. Villmann et al. (Eds.): WSOM+2024, LNNS 1087, pp. 141–151, 2024.
https://doi.org/10.1007/978-3-031-67159-3_17

In this paper, we generalize beyond this prior stream of work by providing a general framework for fairness in partition-based models, a class of models that includes (but is not limited to) decision trees. We also provide a new implementation of fair partition-based machine learning for learning vector quantization (LVQ) models, thus demonstrating the generality of our proposed framework and providing a novel algorithm for fair machine learning that is intrinsically interpretable and compatible with strict resource constraints (constant memory, streaming data, data scarcity). The resulting approach allows for a local and non-linear adaption to fairness constraints. In particular, our method is able to outperform methods based on feature selection and iterative null-space projections which are important state-of-the-art methods in fair machine learning but in the case of composed datasets [9,13].

This paper is structured as follows: First (Sect. 2), we recall the fundamental setup of fairness, focusing on group fairness. We then recall the structure of LVQ models (Sect. 2.2) and briefly discuss the related work on fair machine learning models (Sect. 2.3). Afterward, we introduce our framework for fair partition-based models (Sect. 3.1), from which we derive a fair version of LVQ (Sect. 3.3) by viewing it through the lens of Hebbian learning (Sect. 3.2). We empirically evaluate our method (Sect. 4) on synthetic and real-world data and conclude this work (Sect. 5).

2 Problem Setup

In this section, we recall the setup of fair machine learning focusing on group fairness as well as the vanilla LVQ models and give a short overview of the related work.

2.1 Fair Machine Learning

In supervised machine learning, one commonly considers models $\mathrm{h} : \mathcal{X} \to \mathcal{Y}$ that map samples $\boldsymbol{x}$ in the data space $\mathcal{X}$ to a target or label y in the target space $\mathcal{Y}$. Evaluating a model is done by applying a loss function $\ell : \mathcal{Y} \times \mathcal{Y} \to \mathbb{R}$, e.g. 0-1-loss for classification, MSE for regression, etc., to compare the model output to the desired target $\mathcal{L}_{\mathcal{D}}(\mathrm{h}) = \frac{1}{n} \sum_{i=1}^{n} \ell(\mathrm{h}(\boldsymbol{x}_i), y_i)$ for a data set $\mathcal{D} = \{(\boldsymbol{x}_i, y_i) \in \mathcal{X} \times \mathcal{Y} \mid i = 1, \ldots, n\}$. Training is usually done by optimizing the model parameters such that the loss is minimized. In general, this procedure can lead to models that exhibit undesired biases or unexpected drastic changes in the output which are considered as unfair.

In fair machine learning, the training objective is modified to cope with such problems. Depending on the specific notion of fairness, different objectives are used: *Individual fairness* requires that similar data points are assigned similar outputs. Here, the notion of similarity is usually hand-defined by a human operator based on predefined rules [10].

In this work, we will mainly focus on *group fairness* which extends the label data pair by a *protected* or *sensitive attribute* $(\boldsymbol{x}, y, s)$, $s \in \mathcal{S}$, e.g., sex, race, age, etc. In this case, the training objective is modified to reduce biases that can be linked to the protected attribute which can be evaluated using different scores: *Statistical Parity* describes the idea that all protected groups should have the same amount of *favourable* decisions [10]. *Equal Opportunity* on the other hand focuses only on the cases where the true label is favourable as well [10].

In general, fairness and accuracy cannot be expected to both be optimal, a phenomenon referred to as *fairness-accuracy-tradeoff*. Moreover, different measures are incompatible with each other [10] and some can be full-filled in discriminatory settings [2]. This leads to the problem of choosing an appropriate measure of fairness.

2.2 Learning Vector Quantization Models

In this contribution, we focus on partition-based classifiers, particularly, on Learning Vector Quantization, as intuitive and distance-based classifiers which are designed to solve the multi-classification problem for real vectorial data, i.e., $\mathcal{X} \subset \mathbb{R}^d$ and $\mathcal{Y} = \{1, \ldots, c\}$. They are defined as follows: prototypes $\boldsymbol{w}_1, \ldots, \boldsymbol{w}_P$ with fixed labels $\mathrm{c}(\boldsymbol{w}_j) \in \mathcal{Y}$ are specified such that a good classification and representation of the data is achieved. A new sample $\boldsymbol{x}$ is classified by the winner-takes-all scheme:

$$\boldsymbol{x} \mapsto \mathrm{c}(\boldsymbol{w}_{J(\boldsymbol{x})}) \quad \text{where} \quad J(\boldsymbol{x}) := \operatorname*{arg\,min}_{j \in \{1, \ldots, P\}} d(\boldsymbol{x}, \boldsymbol{w}_j), \tag{1}$$

where $d(\boldsymbol{x}, \boldsymbol{w}_j) = (\boldsymbol{x} - \boldsymbol{w}_j)^\top (\boldsymbol{x} - \boldsymbol{w}_j)$ denotes the squared Euclidean metric.

Given a data set $\mathcal{D}$, prototypes $\boldsymbol{w}_j$ are adapted to minimize the classification error. In this work, we focus on GLVQ which is based on margin maximization [14,15]. In this case, the cost function is given by

$$E = \sum_{i=1}^{n} \Phi\left(\mu(\boldsymbol{x}_i)\right), \tag{2}$$

$$\mu(\boldsymbol{x}_i) = \frac{d(\boldsymbol{x}_i, \boldsymbol{w}^+) - d(\boldsymbol{x}_i, \boldsymbol{w}^-)}{d(\boldsymbol{x}_i, \boldsymbol{w}^+) + d(\boldsymbol{x}_i, \boldsymbol{w}^-)} \tag{3}$$

where Φ is a monotonic increasing function, e. g. the logistic function or the identity $\Phi(\boldsymbol{x}) = \boldsymbol{x}$, $\boldsymbol{w}^+$ denotes the prototype closest to $\boldsymbol{x}_i$ belonging to the same class as $\boldsymbol{x}_i$, i.e., $\mathrm{c}(\boldsymbol{w}^+) = y_i$, and the closest prototype $\boldsymbol{w}^-$ of a different class, i.e., $\mathrm{c}(\boldsymbol{w}^-) \neq y_i$. The term $\mu(\boldsymbol{x}_i) \in [-1, 1]$ is the relative distance difference between the closest correct and incorrect prototype. Depending on the sign a point is either correctly ($\mu(\boldsymbol{x}_i) < 0$), incorrectly classified (> 0), or on the decision boundary ($= 0$).

Training takes place based on a given training set as above and consists of two steps. First, the prototypes are initialized, typically this is either done within the class centers or by using the cluster centers of a clustering algorithm. Second, the cost function E is minimized using a gradient descent scheme. In this work, we choose Φ as the swish activation function [18].

2.3 Related Work

There is a rich body of literature on fair machine learning. According to [10], creating fair models can be split into three main segments: 1) *pre-processing* the dataset reduces biases before training, 2) controlling the bias during training of the model using a suited objective (*in-processing*), and 3) *post-processing* the trained model to remove potentially learned biases.

Common pre-processing approaches transform the data by removing the protected attribute or projecting onto an orthogonal subspace (*null-space*) [13]. This process can be iterated multiple times. We refer to this method as "iterative null-space projection" (INP). However, such methods can remove valuable information while information on the protected attribute can still be contained in (highly non-linear) correlations [10,13].

A simple in-processing technique is to include the desired fairness criterion as a loss term or constraint during training [5,20].

Adversarial Debiasing [19] is a state-of-the-art in-process technique that uses ideas similar to [13]: A deep representation that is well suited for the learning problem but unsuited for predicting the protected attributed is learned using a kind of adversarial training. A disadvantage of this approach is the instability of adversarial learning [19].

Several approaches exist to train fair decision trees [4,8,12]. A particular simple in-processing technique is proposed by [4]. Here, the decrease in impurity of the protected attribute is considered as a penalty in the splitting criterion.

Although rich literature exists on partition-based models, to the best of our knowledge, there are no previous works on fairness of LVQ models.

3 Method

In this section, we first present a general method for training fair partition-based models. We then apply this scheme to GLVQ to derive a novel, fair LVQ variant.

3.1 Fairness for General Partition-Based Models

The general concept of partition-based models relies on partitioning the data space into regions that are homogeneous w.r.t to the targets. Each region is targeted using a single, usually simple model. We will focus on constant models, the resulting partition-based models are thus given by $\mathrm{h}(\boldsymbol{x}) = \mathrm{c} \circ \mathrm{p}(\boldsymbol{x})$ where $\mathrm{p} : \mathcal{X} \to \mathbb{N}$ is a partition and $\mathrm{c} : \mathbb{N} \to \mathcal{Y}$. Notice that both decision trees as well as LVQ belong to this type of model. In the case of decision trees, the regions are given by the leaves; in the case of LVQ, they are induced by the prototypes.

Given a dataset $\mathcal{D}$, partition-based models can be trained by minimizing the loss over p and c as usual. While optimizing p may be difficult, optimizing c given p is usually simple, e.g. the average in the partition for decision trees.

The main problem is thus to find the right partition so that the regions are homogeneous with respect to the target. To ensure fairness, we use the opposite criterion for the protected attribute, i.e., we reward a partition for a low loss on predicting the targets (*local homogeneity*) but also for a high loss for predicting the protected attribute (*local heterogeneity*). We achieve this by adding a regularization term, weighted by a scaling constant C, yielding a new optimization problem:

$$\min_{\mathrm{p}} \left[\min_{\mathrm{c}} \mathcal{L}_{\mathcal{D}}^{\mathrm{class}}(\mathrm{c} \circ \mathrm{p}) - C \cdot \min_{\mathrm{f}} \mathcal{L}_{\mathcal{D}}^{\mathrm{fair}}(\mathrm{f} \circ \mathrm{p}) \right], \tag{4}$$

where $\mathcal{L}_{\mathcal{D}}^{\mathrm{class}}$ and $\mathcal{L}_{\mathcal{D}}^{\mathrm{fair}}$ are the empirical losses concerning the target and protected attribute, respectively. We emphasize that this formulation is *independent of a specific fairness metric* as we generally punish good classification w.r.t. the protected attribute. It also directly *applies to regression* for both target and protected attributes.

In the next step, we discuss our training scheme from a Hebbian perspective before applying it to LVQ.

3.2 Fair Hebbian Learning, Pseudo-classes, and FairGLVQ

The fundamental idea of Hebbian learning is to reinforce or weaken the connection between neurons that are active at the same time. In the case of classical LVQ, this is realized by moving the prototypes closer to or further away from the observed data point. To assure fairness similar to Eq. (4), we also want to have an inhibiting effect if a neuron is strongly associated with the protected attribute. For LVQ, this means pushing the prototype away. This effect can easily be obtained by assigning a second *pseudo-class* representing the majority protected attribute (cf. Eq. (4)) to each prototype that is used with an opposite gradient during training. Notice that the pseudo-class is updated during the training process. We will discuss more technical details in the next section where we give an explicit algorithm as an extension of GLVQ: FairGLVQ.

3.3 Algorithm Design

This section describes the technical details of how our algorithm is designed to implement the general method presented in Sect. 3.1. Following the ideas from the last section, we adapt Eq. (2) to punish good classification w.r.t. the pseudo-classes. Let $\mu_{\mathrm{fair}}(\boldsymbol{x}_i)$ denote Eq. (3) for the pseudo-classes. Then, the new FairGLVQ cost function is given as:

$$E_{\mathrm{fair}} = \sum_{i=1}^{n} \Phi(\mu_{\mathrm{class}}(\boldsymbol{x}_i)) - C \cdot \Phi(\mu_{\mathrm{fair}}(\boldsymbol{x}_i)). \tag{5}$$

The parameter C again controls the strength of the regularization. The resulting algorithm is presented in Algorithm 1. Observe that up to lines 8 to 14 this is just a mini-batch version of the standard GLVQ training algorithm. Here we use a mini-batch approach as the computation of the fair loss requires the pseudo-classes which are expensive to compute.

Algorithm 1. FairGLVQ

1: **Input:** $\mathcal{D}$ train data, N number of epochs, M batch-size, η learning rate, regularization parameter C, missing pseudo-class parameter α
2: Initialize prototypes $\boldsymbol{W} \leftarrow \{\boldsymbol{w}_1, ..., \boldsymbol{w}_P\}$ ▷ Use fairness-adapted clustering
3: **for** N steps **do**
4: $\quad \mathbf{G}^{\text{class}} \leftarrow (0, \ldots, 0)$, $\mathbf{G}^{\text{fair}} \leftarrow (0, \ldots, 0)$, $n_{\text{fair}} \leftarrow 0$, $n_{\text{class}} \leftarrow 0$
5: $\quad$ Pick mini-batch $B = \{(\boldsymbol{x}_1, y_1, s_1), (\boldsymbol{x}_2, y_2, s_2), ..., (\boldsymbol{x}_M, y_M, s_M)\} \subset \mathcal{D}$
6: $\quad$ **for all** $(\boldsymbol{x}, y, s) \in B$ **do**
7: $\qquad \boldsymbol{w}^{\pm}_{\substack{\text{class}\\\text{fair}}} \leftarrow \arg\min_{\boldsymbol{w} \in \boldsymbol{W}^{\pm}_{\substack{\text{class}\\\text{fair}}}} d(\boldsymbol{x}, \boldsymbol{w})$ ▷ Determine $\boldsymbol{w}^+_{\text{class}}, \boldsymbol{w}^-_{\text{class}}, \boldsymbol{w}^+_{\text{fair}}, \boldsymbol{w}^-_{\text{fair}}$
8: $\qquad$ **if** $\boldsymbol{w}^-_{\text{fair}}$ does not exist **then**
9: $\qquad\quad d(\boldsymbol{x}_i, \boldsymbol{w}^-_{\text{fair}}) := \frac{1}{\alpha} \cdot d(\boldsymbol{x}_i, \boldsymbol{w}^+_{\text{fair}})$, $\boldsymbol{W}_{\text{update}} \leftarrow \{\boldsymbol{w}^+_{\text{fair}}, \boldsymbol{w}^+_{\text{class}}, \boldsymbol{w}^-_{\text{class}}\}$
10: $\qquad$ **else if** $\boldsymbol{w}^+_{\text{fair}}$ does not exist **then**
11: $\qquad\quad d(\boldsymbol{x}_i, \boldsymbol{w}^+_{\text{fair}}) := \alpha \cdot d(\boldsymbol{x}_i, \boldsymbol{w}^-_{\text{fair}})$, $\boldsymbol{W}_{\text{update}} \leftarrow \{\boldsymbol{w}^-_{\text{fair}}, \boldsymbol{w}^+_{\text{class}}, \boldsymbol{w}^-_{\text{class}}\}$
12: $\qquad$ **else**
13: $\qquad\quad \boldsymbol{W}_{\text{update}} \leftarrow \{\boldsymbol{w}^+_{\text{fair}}, \boldsymbol{w}^-_{\text{fair}}, \boldsymbol{w}^+_{\text{class}}, \boldsymbol{w}^-_{\text{class}}\}$
14: $\qquad$ **end if**
15: $\qquad \mathbf{G}^{\substack{\text{class}\\\text{fair}}}_{\boldsymbol{w}} \leftarrow \mathbf{G}^{\substack{\text{class}\\\text{fair}}}_{\boldsymbol{w}} + \nabla_{\boldsymbol{w}} E_{\text{fair}}$ for $\boldsymbol{w} \in \boldsymbol{W}_{\text{update}}$ ▷ Compute gradient
16: $\qquad n_{\text{fair}} \leftarrow n_{\text{fair}} + |\boldsymbol{W}_{\text{update}}| - 2$, $n_{\text{class}} \leftarrow n_{\text{class}} + 2$
17: $\quad$ **end for**
18: $\quad$ **for all** $\boldsymbol{w} \in \boldsymbol{W}$ **do** ▷ Update prototypes using mini-batch gradients $\mathbf{G}^{\substack{\text{class}\\\text{fair}}}_{\boldsymbol{w}}$
19: $\qquad \boldsymbol{w} \leftarrow \boldsymbol{w} - \eta \left(\frac{1}{n_{\text{class}}} \mathbf{G}^{\text{class}}_{\boldsymbol{w}} + \frac{1}{n_{\text{fair}}} \mathbf{G}^{\text{fair}}_{\boldsymbol{w}} \right)$
20: $\quad$ **end for**
21: $\quad$ **for all** $\boldsymbol{w}_j \in \boldsymbol{W}$ **do** ▷ Update pseudo-classes using majority vote
22: $\qquad s(\boldsymbol{w}_j) \leftarrow \arg\max_{s \in \mathcal{S}} |\{(\boldsymbol{x}, y, s) \in \mathcal{D} \mid s = s(\boldsymbol{w}_j) \wedge J(\boldsymbol{x}) = j\}|$
23: $\quad$ **end for**
24: **end for**

As the pseudo-classes are not fixed but reassigned according to the majority vote, we have to take two edge cases into account.

Edge-Case 1: empty receptive field: we choose a pseudo-class at random.

Edge-Case 2: no (in-)correct pseudo-class: computing μ_{fair} requires the closest prototype with the same/different pseudo-class which may not exist. In this case, we define $d(\boldsymbol{x}, \boldsymbol{w}^-_{\text{fair}}) := \frac{1}{\alpha} \cdot d(\boldsymbol{x}, \boldsymbol{w}^+_{\text{fair}})$ and $d(\boldsymbol{x}, \boldsymbol{w}^+_{\text{fair}}) := \alpha \cdot d(\boldsymbol{x}, \boldsymbol{w}^-_{\text{fair}})$ if $\boldsymbol{w}^-_{\text{fair}}$ or $\boldsymbol{w}^+_{\text{fair}}$ does not exist, respectively, which corresponds to simulating a prototype such that $\boldsymbol{x}$ is considered to have a protected attribute different from the overall closest prototype (lines 8–14 in Algorithm 1). This setting implies $\mu_{\text{fair}}(\boldsymbol{x}_i) = \frac{\alpha-1}{\alpha+1} \in (0, 1)$. Therefore, the parameter $\alpha \in (1, \infty)$ determines the value of the fairness penalty in this case. We use $\alpha = 2$ as a default value.

In contrast to the usual LVQ initialization which is class-specific, we initialize prototypes for every class at every found cluster center with a small perturbation.

4 Experiments

In this section, we empirically evaluate our approach comparing it to other methods from the literature. We consider both, a synthetic dataset to showcase the

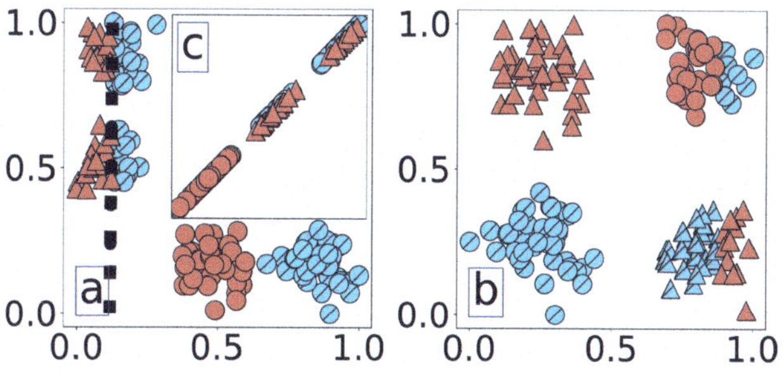

Fig. 1. Synthetic datasets (*a*: "local" with linear model (black), *b*: "XOR", *c*: fair projection of "local"). Label is color/hatched; Protected attribute is shape.

Table 1. Mean (and standard deviation) Accuracy (Acc; higher is better) and Statistical Parity (SP; lower is better) on synthetic datasets over a 5-fold cross-validation.

		FairGLVQ (ours)	INP	GLVQ	const.
Local	Acc	0.76(±0.02)	0.51(±0.01)	0.99(±0.00)	0.51(±0.00)
	SP	0.12(±0.13)	0.24(±0.12)	0.65(±0.01)	0.00(±0.00)
XOR	Acc	0.88(±0.01)	0.84(±0.05)	0.99(±0.00)	0.50(±0.00)
	SP	0.03(±0.02)	0.24(±0.12)	0.23(±0.04)	0.00(±0.00)

specific properties of FairGLVQ as well as two real-world datasets to demonstrate the practical relevance of our approach.

For both experiments, we make use of three baselines: GLVQ, GLVQ with linear pre-processing (INP), and a constant baseline model. For GLVQ, we initialize the prototypes based on a per-class clustering using k-means, otherwise with the same parameters as we use for FairGLVQ. For the linear pre-processing, we use the iterative null-space projection method (INP) [13] and apply standard GLVQ to the pre-processed data. Notice that for $C = 0$ or no projection FairGLVQ and INP are equal to GLVQ (up to prototype initialization).

4.1 Synthetic Data

We start by showcasing the advantages of FairGLVQ on synthetic data. We consider two datasets with 2 classes and a binary protected attribute (See Fig. 1 for a visualization). "XOR" consists of 4 Gaussian's, the occurrence of classes and protected values can be described by xors shifted against each other along one axis. "local" consists of two separate classification problems, one aligned with the protected attribute, the other unaffected by the protected attribute. The protected attribute is not considered as part of the data.

We use 4 prototypes per class for XOR and 5 for local. As the dataset is two-dimensional we project onto one dimension for INP. For FairGLVQ, we use

$M = 250$, $\alpha = 2$, $\eta = 0.005$, and $C = 1.5$ for local and $C = 1.25$ for XOR. We train for $N = 250$ epochs. We run a 5-fold cross-validation and measure Accuracy and Statistical Parity (the latter as the absolute difference). The results are shown in Table 1. As can be seen, GLVQ has the highest accuracy but is very unfair. FairGLVQ outperforms INP with regard to accuracy and shows comparable fairness. For the "local" dataset, this can be explained by observing that there is no fair projection that still provides sufficient information for classification as illustrated in Fig. 1. For the "XOR" dataset, a linear classifier cannot reliably learn a useful projection dimension. This shows the advantage of the local and non-linear fairness criterion considered by FairGLVQ.

4.2 Real-World Benchmarks

Besides theoretical datasets, we consider the real-world datasets "COMPAS" [1] (protected attribute race) and "Adult Income" [6] (Adult; protected attribute sex). In both cases, the protected attribute is provided as a feature.

For both datasets, we used 20 prototypes per class, and 500 epochs with a learning rate of 0.05. For FairGLVQ, we used $\alpha = 2$, $M = 200$ for COMPAS, and $M = 1000$ for Adult. We considered different regularization strengths, i.e., C for FairGLVQ and the number of projection dimensions for INP.

We document Accuracy, Statistical Parity, and Equal Opportunity (the latter two as absolute difference) using a 5-fold cross-validation. The results are shown in Fig. 2. As shown, both algorithms are finding good compromises within the fairness-accuracy-tradeoff for the COMPAS dataset and perform very similarly with slight differences in high and low regularization tradeoffs respectively. For Adult, INP is only able to improve fairness up to a certain point before Accuracy drops to constant baseline level. One explanation for this effect is that the projection removes so much information that classification is no longer possible, similar to the "local" data set. In particular, FairGLVQ outperforms INP on Adult in case of high fairness. Conversely, this implies that COMPAS admits a linear decomposition into subspaces relevant for predicting the label and protected attribute, respectively.

Note that both approaches improve in both fairness metrics. However, for the Adult dataset the regularization does not necessarily lead to a monotonic decline for weak regularization in both algorithms and for strong regularization in FairGLVQ. Otherwise, the relation is monotonic as expected.

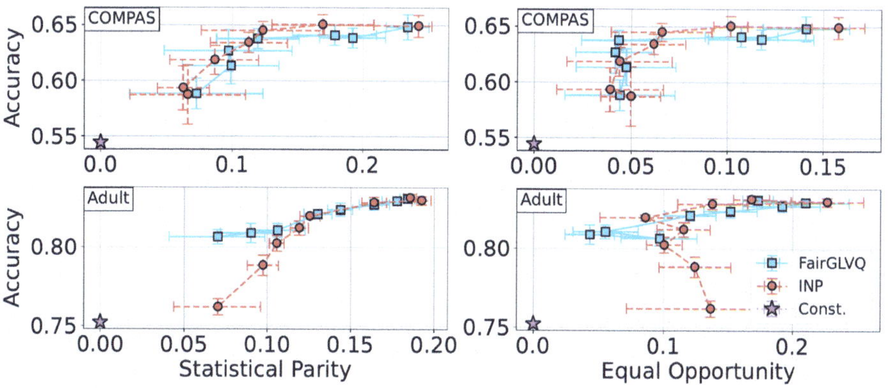

Fig. 2. Empirical evaluation of real-world datasets for different regularizations using 5-fold cross-validation. Fairness score (x-axis, lower is better) and Accuracy (y-axis, higher is better). COMPAS (top) and Adult (bottom).

5 Conclusion and Discussion

In this paper, we considered the problem of group fairness for partition-based models with a particular focus on LVQ. We derived a general framework to impose fairness in partition-based models via an intuitive notion of fairness (punishing good classification w.r.t. protected attribute) that is not limited to discrete protected attributes nor specific fairness metrics like Statistical Parity or Equal Opportunity.

Based thereon, we derived a new, fair version of the classical LVQ models by considering the framework through the lens of Hebbian learning. We therefore believe that our considerations can also be applied to other models and methods. In particular, the application to other LVQ models like GMLVQ [15], LGMLVQ [15] or KGLVQ [11] is subject to further work.

We evaluated FairGLVQ, showing a good fairness-accuracy-tradeoff. In particular, we showed that it can outperform global approaches that are based on iterated null-space projection which is a state-of-the-art method for both fairness and federated learning. Probing the potential of FairGLVQ for handling composed datasets is yet another interesting line of further work as well as investigating the performance on multi-class/multi-protected problems. Due to the locality of our approach, an investigation into individual fairness is another interesting frontier. Our code is available online[1].

Acknowledgments. We gratefully acknowledge funding from the European Research Council (ERC) under the ERC Synergy Grant Water-Futures (Grant agreement No. 951424).

[1] https://github.com/Felix-St/FairGLVQ.

References

1. Angwin, J., Larson, J., Mattu, S., Kirchner, L.: Machine bias. https://www.propublica.org/article/machine-bias-risk-assessments-in-criminal-sentencing
2. Dwork, C., Hardt, M., Pitassi, T., Reingold, O., Zemel, R.: Fairness through awareness, pp. 214–226. Association for Computing Machinery, New York (2012)
3. European Commission and Directorate-General for Communications Networks, Content and Technology: Ethics guidelines for trustworthy AI (2019)
4. Kamiran, F., Calders, T., Pechenizkiy, M.: Discrimination aware decision tree learning. In: ICDM 2010, The 10th IEEE International Conference on Data Mining, pp. 869–874. IEEE Computer Society (2010)
5. Kamishima, T., Akaho, S., Asoh, H., Sakuma, J.: Fairness-aware classifier with prejudice remover regularizer. In: Flach, P.A., De Bie, T., Cristianini, N. (eds.) ECML PKDD 2012. LNCS (LNAI), vol. 7524, pp. 35–50. Springer, Heidelberg (2012). https://doi.org/10.1007/978-3-642-33486-3_3
6. Kohavi, R.: Census Income. UCI Machine Learning Repository (1996). https://doi.org/10.24432/C5GP7S
7. Laux, J., Wachter, S., Mittelstadt, B.: Trustworthy artificial intelligence and the European union AI act: on the conflation of trustworthiness and acceptability of risk. Regul. Gov. **18**(1), 3–32 (2023)
8. van der Linden, J.G.M., de Weerdt, M., Demirovic, E.: Fair and optimal decision trees: a dynamic programming approach. In: NeurIPS (2022)
9. Lövdal, S., Biehl, M.: Improved interpretation of feature relevances: iterated relevance matrix analysis (IRMA). In: ESANN 2023 Proceedings, pp. 59–64 (Oct 2023)
10. Mehrabi, N., Morstatter, F., Saxena, N., Lerman, K., Galstyan, A.: A survey on bias and fairness in machine learning. ACM Comput. Surv. **54**(6), 1–35 (2021)
11. Qin, A., Suganthan, P.: A novel kernel prototype-based learning algorithm. In: Proceedings of the 17th International Conference on Pattern Recognition, ICPR 2004, vol. 4, pp. 621–624 (2004)
12. Ranzato, F., Urban, C., Zanella, M.: Fairness-aware training of decision trees by abstract interpretation. In: CIKM 2021, Queensland, Australia, 1–5 November 2021, pp. 1508–1517. ACM (2021)
13. Ravfogel, S., Elazar, Y., Gonen, H., Twiton, M., Goldberg, Y.: Null it out: guarding protected attributes by iterative nullspace projection. In: Proceedings of the 58th Annual Meeting of the Association for Computational Linguistics, pp. 7237–7256. Association for Computational Linguistics (2020)
14. Sato, A., Yamada, K.: Generalized learning vector quantization. In: Advances in Neural Information Processing Systems 8, NIPS, Denver, CO, USA, 27–30 November 1995, pp. 423–429. MIT Press (1995)
15. Schneider, P., Biehl, M., Hammer, B.: Adaptive relevance matrices in learning vector quantization. Neural Comput. **21**(12), 3532–3561 (2009)
16. Strotherm, J., Hammer, B.: Fairness-enhancing ensemble classification in water distribution networks. In: Rojas, I., Joya, G., Catala, A. (eds.) IWANN 2023. LNCS, vol. 14134, pp. 119–133. Springer, Cham (2023). https://doi.org/10.1007/978-3-031-43085-5_10
17. Strotherm, J., Müller, A., Hammer, B., Paaßen, B.: Fairness in KI-Systemen. arXiv arXiv:2307.08486 (2023). in press at Springer, German
18. Villmann, T., Ravichandran, J., Villmann, A., Nebel, D., Kaden, M.: Investigation of activation functions for generalized learning vector quantization. In: Vellido, A., Gibert, K., Angulo, C., Martín Guerrero, J.D. (eds.) WSOM 2019. AISC,

vol. 976, pp. 179–188. Springer, Cham (2020). https://doi.org/10.1007/978-3-030-19642-4_18
19. Zhang, B.H., Lemoine, B., Mitchell, M.: Mitigating unwanted biases with adversarial learning. In: Proceedings of the 2018 AAAI/ACM Conference on AI, Ethics, and Society, pp. 335–340. ACM (2018)
20. Zhang, W., Ntoutsi, E.: FAHT: an adaptive fairness-aware decision tree classifier. In: Proceedings of the Twenty-Eighth International Joint Conference on Artificial Intelligence, IJCAI 2019, Macao, China, 10–16 August 2019, pp. 1480–1486. ijcai.org (2019)

About Interpretable Learning Rules for Vector Quantizers - A Methodological Approach

Ronny Schubert(✉) and Thomas Villmann

Saxon Institute for Computational Intelligence and Machine Learning, University of Applied Sciences Mittweida, Mittweida, Germany
schuber6@hs-mittweida.de

Abstract. In this work, we will investigate two approaches to deploy learning rules. A combination of these approaches is used to create a generic learning rule for prototype-based models with the emphasis on interpretability. In this regard, we will show how the learning rules are associated to the underlying decision making of such models. Moreover, the work concludes by giving possible interpretations of these rules and anchor points for developing related explanations and designing comprehensible learning rules.

1 Introduction

In the last years, a great effort was put into creating explanations for the decision making of machine learning models [16,21]. However, although unsupervised *vector quantization* (**VQ**) and supervised *learning*-VQ (**LVQ**) models are often claimed to be *ante-hoc* interpretable for representation and classification learning, according explanations mostly rely on *post-hoc* methods like *counterfactual explanations* [1] or *feature contributions/correlations* [23]. In this regard, discussions about the completeness of such explanations may arise as well as about the actual border between *ante-hoc* interpretability and explainable artificial intelligence when relying on *post-hoc* explanations [7,16,24]. However, the scope of this work is not to resolve this discussion, but to inspire how the attribute of *ante-hoc* could be used to derive explanations for the decision making of VQ models. In this regard, we will consider two approaches to derive learning rules and relate these to VQ models by using the decision making itself.

2 Oja's Learning Rule - A Input-Output Relation View

To create an interpretable learning rule, an important early approach was proposed by [20] for unsupervised linear perceptron learning known as *Oja's Rule.*

R. S. is supported by grants of the Agriculture Ministry of Germany, funding code: 28DK112B20.

T. Villmann et al. (Eds.): WSOM+2024, LNNS 1087, pp. 152–162, 2024.
https://doi.org/10.1007/978-3-031-67159-3_18

Given a data sample $\boldsymbol{x}_i \in \mathcal{X} \subseteq \mathbb{R}^n$, the rule is derived as a consequence of the input-output relation

$$\nu_{Oja}(\boldsymbol{x}_i) = \langle \boldsymbol{x}_i, \boldsymbol{w} \rangle = \sum_j x_i^j \cdot w^j \tag{1}$$

known as the Hebb-excitation [11] with weight vector $\boldsymbol{w}$ and an additional normalization constraint for $\boldsymbol{w}$. More precisely, the resulting perceptron response $\nu_{Oja}(\boldsymbol{x})$ determines the Hebbian learning dynamic for a stochastically drawn sample $\boldsymbol{x}_i$ presented at time t by

$$\boldsymbol{w}(t+1) = \boldsymbol{w}(t) + \epsilon \cdot \nu_{Oja}(\boldsymbol{x}_i) \cdot [\boldsymbol{x}_i - \nu_{Oja}(\boldsymbol{x}_i) \cdot \boldsymbol{w}(t)] + \mathcal{O}(\epsilon^2) \tag{2}$$

where $0 < \epsilon \ll 1$ is the learning rate. The term $\nu_{Oja}(\boldsymbol{x}_i) \cdot \boldsymbol{x}_i$ is representing the Hebb-learning, whereas $(\nu_{Oja}(\boldsymbol{x}_i))^2 \cdot \boldsymbol{w}$ realizes the normalization constraint. Moreover, in [20] it was shown that this learning converges the weight vector $\boldsymbol{w}$ to be the first principal component of the input covariance matrix in case of centralized data. In consequence, Oja's rule constitutes a self-organizing behavior, since there were no assumptions made about the actual objective of the relation (1), i. e. which result should be achieved. Further, the normalization term can be interpreted in a biological sense as a *forgetting* of the Hebbian learning and, thus, constitutes a comprehensible behavior. For the rest of this work we will refer to (2) and the respective context as *Oja's view* with a general function ν_{Oja} determining the learning dynamic, i. e. consideration of a learning rule without a regarding objective.

3 Competition Functions - An Objective View for Vector Quantization

Apart from Oja's rule, generalized learning rules for prototype-based vector quantization networks have also been developed, but without the emphasis on interpretable behavior. Accordingly, in [14], a general learning rule for prototypes $\boldsymbol{p}_i \in \mathcal{P} = \{\boldsymbol{p}_1, \ldots, \boldsymbol{p}_c\} \subset \mathbb{R}^n$ was established as

$$\Delta \boldsymbol{p}_i = \epsilon \cdot \sum_j \alpha_{ij}(\boldsymbol{x}_j - \boldsymbol{p}_i(t)) \tag{3}$$

where $(\boldsymbol{x}_j - \boldsymbol{p}_i(t))$ is a vector shift corresponding to a gradient of the squared Euclidean distance $||\boldsymbol{x}_j - \boldsymbol{p}_i(t)||^2$ with respect to the prototype vector $\boldsymbol{p}_i(t)$. The so-called *competition functions* α_{ij} play an essential role in this approach and need to fulfill several axioms in dependence on the resulting vector quantizer model. In particular, the learning dynamic was proposed to realize several fuzzy-clustering algorithms using the general objective

$$R_m = \frac{1}{M} \sum_j \left(\frac{1}{c} \sum_l \left(||\boldsymbol{x}_j - \boldsymbol{p}_l||^2 \right)^{\frac{1}{1-m}} \right)^{1-m} \tag{4}$$

depending on the competition functions α_{ij} and the parameters m, $M = |\hat{\mathcal{X}}|$ with $\hat{\mathcal{X}} \subseteq \mathcal{X}$ and $c = |\mathcal{P}|$. Nevertheless, although the presented model of [14] comes with a mathematical theory, corresponding interpretations as well as explanations of the decision making rely on a considered variant, i.e. they depend on the choice of the competition functions determining the objective. As from now, we will refer to (3) derived by considering objective functions as *Objective View* and use the short-hand notation $\boldsymbol{p}_i = \boldsymbol{p}_i(t)$.

4 A Learning Rule for Vector Quantization

Oja's view yields that an objective does not necessarily need to be considered in the first place when deriving a learning rule, while the *Objective View* helps to generate general prototype-based learning rules. A combination of these approaches results in a generic learning rule for VQ which is sufficiently general to also represent rules that were not derived from an objective function. We therefore consider an update scheme like

$$\Delta \boldsymbol{p}_i = \sigma \cdot \epsilon \cdot \sum_j g_{ij} \cdot \boldsymbol{s}_{ij} \tag{5}$$

which we will refer to as *generic rule*. Here, $\sigma \in \{-1, 1\}$ represents the direction of the update (*sign*), which later is considered in relation to the update method, whereas the functions g_{ij} and $\boldsymbol{s}_{ij}$ are denoted as *gain* and (abstract) *shift*, respectively.

As for the scope of this work, the explanations of a VQ model using the generic rule (5) particularly consider these functions, i.e. their respective influence on the full update prototype scheme. Obviously, to create an explainable mechanism, both gain and shift must be meaningful to a certain extent, i.e. the construction of g_{ij} and $\boldsymbol{s}_{ij}$ must be comprehensible and transparent.

A similar general approach was considered in [3], but with the focus to analyze the LVQ learning dynamics for several LVQ variants. In contrast, here we will focus on the design of appropriate shift and gain in terms of Oja's and Objective View to achieve interpretable models.

4.1 The Generic Rule in the Objective View

Applying the ideas of the Objective View to the generic learning rule in (5), the update can be derived as a gradient in terms of a differentiable objective function O:

$$\Delta \boldsymbol{p}_i = \sigma \cdot \epsilon \cdot \sum_j \frac{\partial O}{\partial b_i(\boldsymbol{x}_j)} \cdot \frac{\partial b_i(\boldsymbol{x}_j)}{\partial \boldsymbol{p}_i} \tag{6}$$

with the gain functions $g_{ij} = \frac{\partial O}{\partial b_i(\boldsymbol{x}_j)}$ and the shifts $\boldsymbol{s}_{ij} = \frac{\partial b_i(\boldsymbol{x}_j)}{\partial \boldsymbol{p}_i}$. Here, the function $b_i(\boldsymbol{x})$ constitutes the *belongingness* of prototype $\boldsymbol{p}_i$ to the sample $\boldsymbol{x}$, which is commonly modeled as a (dis-)similarity [22]. For example, if the squared Euclidean metric is used as in (4), the abstract shift appears to be an ordinary vector shift. In divergence-based vector quantization as investigated in [25], the abstract shift is obtained as the respective gradient of the divergence measure in use with respect to the prototype $\boldsymbol{p}_i$.

Generally, for the Objective View, the belongingness function is required to be differentiable at least. Moreover, as pointed out above, σ specifies the update type, i.e. gradient ascent learning for the objective O in case of $\sigma = +1$ and descent learning for $\sigma = -1$ and hence relates to the formulation of the objective O. Furthermore, assuming the objective O is of the form (4), we are able to consider the gain functions g_{ij} in (6) in the context of the competition functions α_{ij} in (3), supposing the respective axioms declared in [14] are fulfilled. Anyway, while these axioms would imply $g_{ij} \geq 0$ given $\sigma = -1$, we here also allow $g_{ij} < 0$, such that g_{ij} accounts for both rewarding and punishing updates, since this is an important requirement for more general supervised VQ scenarios.

According to this interpretation, the gain g_{ij} in the Objective View can be seen as to scale and direct the shift in terms of the objective function O and the interplay with σ. This means in detail, that the gain may weight the information for the prototype adaptation (strength) and, for rivaling scenarios, may also penalize, as required for reward-punishment based update schemes. In consequence, the gain of (6) in the Objective View expresses the influence of a prototype $\boldsymbol{p}_i$ on the objective in terms of its belongingness $b_i(\boldsymbol{x}_j)$ to sample $\boldsymbol{x}_j$ and, hence, is related to the specific decision making in use of a VQ scheme, which will be considered in Sect. 4.2 more deeply.

The taken perspective and resulting insights in this section should help to comprehend the learning dynamic of a general VQ model using a learning rule like (6). Yet, to gain a profound understanding of the resulting decision making, i.e. of the inference, explanations would still rely on the investigated variant. Nevertheless, by using Oja's view, we want to enhance the meaning of g_{ij} further, allowing to understand (5) and (6) in terms of the decision making and output generation.

4.2 Decision Making and Output Generation of VQ as an Input-Output Relation

In the following we investigate more deeply the role of the gain g_{ij} with respect to the decision making and subsequent output generation. In particular, first we want to examine this process in terms of an input-output relation, which, in turn, will be necessary for the interpretation of (5) and (6) in Oja's view. In this regard, we consider a model response $\nu_{VQ}(\boldsymbol{x})$ of VQ usually realized by *nearest-prototype principle* (**NPP**) or by its generalized version k-NPP if more than one prototype is considered [16,19]. More precisely, the k-NPP assigns a sample $\boldsymbol{x}$ to its k-nearest neighbors in $\mathcal{P}$ or their respective classes via their belongingness. Describing this process in mapping-notation, we find that $\boldsymbol{x}$ needs firstly to be mapped to $\mathcal{B} = \{b_1(\boldsymbol{x}), \ldots, b_c(\boldsymbol{x})\}$ assigning to each prototype $\boldsymbol{p}_i$ its $b_i(\boldsymbol{x})$ value by means of an abstract function B. Next, the k-nearest prototypes are selected via a decision function D and, lastly, the final output is determined via a specific assignment function A. The whole process generates the network output and hence is formally determined by the input-output relation

$$\nu_{VQ} : \mathcal{X} \xrightarrow{B} \mathcal{B} \xrightarrow{D} \tilde{\mathcal{I}} \xrightarrow{A} \mathcal{A} \tag{7}$$

where $\tilde{\mathcal{I}} \subseteq \mathcal{I} = \{1, \ldots, c\}$ is determined by the decision mapping D and $|\tilde{\mathcal{I}}| = k$. The set $\mathcal{A}$ stands for a general VQ-model output. Note that not all dependencies are mentioned in the short-hand notation (7), since not all arguments used by the respective functions are shown. For example, B also depends on $\mathcal{P}$ and supposing, that $\mathcal{Y} = \{y_1, \ldots, y_c\}$ are the classes y_i associated to their respective prototype $\boldsymbol{p}_i$, then A also depends on $\mathcal{Y}$. In unsupervised VQ, $\mathcal{A}$ may be the indices of the k winning prototypes if the assignment is assumed as the identity $A(\cdot) = Id(\cdot)$. In classification learning A could realize a majority vote or simply the class assignment y_s of the most appropriate prototype according to the belongingness.

Hence, the decision function D simply selects prototypes based on their belongingness. This includes also shallow overall *winner-take-all* (**WTA**) determination, i.e. $D(\mathcal{B}) = \arg\min_{l \in \mathcal{C}} b_l(\boldsymbol{x})$ if $b(\boldsymbol{x})$ constitutes a dissimilarity and, analogously, $D(\mathcal{B}) = \arg\max_{l \in \mathcal{C}} b_l(\boldsymbol{x})$ if $b(\boldsymbol{x})$ yields a similarity with $\mathcal{C} = \{1, \ldots, c\}$.

Taking this mapping scheme (7) as formalization of the input-output relation of the VQ model, we recognize that the decision and the assignment mappings are most important for the inference phase but do not necessarily contribute to the prototype learning.[1] Thus, a potential explanation of this relation relies on the explanations given for B, D and A and their interplay, but do not necessarily cover explanations regarding the learning process or dynamic, which was used to build the knowledge for the models conclusion. Consequently, such a limited explanation would only cover how a VQ model decides, but could not or only partially provide a reasoning to substantiate the conclusion in terms of the model learning dynamic. Thus, to achieve full model credibility, we explicitly have to deal with the explanation of the learning dynamic for the prototypes as well.

4.3 The Generic Rule According to Oja's View

Considering the gain g_{ij} in the Objective View (6), the aforementioned formal descriptions of B, D and A have to represent to a certain extend also the objective function. Consequently, the final decision and assignment mappings should influence the choice of the learning dynamic by means of appropriate learning rules. Yet, heuristic VQ-models like LVQ [15] or cognitive-inspired LVQ [17], may lack a respective objective function. Moreover, it might be impossible to formulate an objective for a given learning dynamic such that a respective learning rule can be derived for optimization as it is known for the original self-organized map

[1] To emphasize, the belongingness function depends on given prototypes but does not requires knowledge about the optimization process for the prototype.

[6]. For those models, explanations of g_{ij} and $\boldsymbol{s}_{ij}$ for the learning dynamic have to be made in agreement with the demanded interpretability of the VQ-model taking Oja's View. However, the explainable learning dynamic determined by the gain g_{ij} and shift $\boldsymbol{s}_{ij}$ has to keep the consistency with the input-output-relation according to the formalism (7) and thus preserve the respective interpretability.

In particular, there might be a difference in decision making during the learning and during the inference phase: For the inference phase realized by ν_{VQ}, decision making influences the later output, while the decision making during learning determines the prototype update strategy as realized for $\boldsymbol{w}$ by means of ν_{Oja} in Oja's Rule. For example, decision making in the inference phase may be the simple WTA decision, whereas *neighborhood-cooperativeness* is taken into account for prototype learning as in self-organizing maps or neural gas [15,18]. For classification learning in LVQ models, the prototype adaptation is following on the one hand side the WTA for both inference and learning, whereas the learning rule also depends on class agreement between the presented training sample $\boldsymbol{x}_j$ and the considered prototype $\boldsymbol{p}_s$ selected by the WTA-rule [2,13,19].

To make this difference better visible we introduce the formal input-output relation scheme

$$\bar{\nu}_{VQ} : \mathcal{X} \xrightarrow{B} \mathcal{B} \xrightarrow{\bar{D}_{\Gamma}} \bar{\mathcal{I}} \xrightarrow{L} \mathcal{L} \tag{8}$$

for the learning process in analogy to (7) for the inference process. Here, the decision function $\bar{D}_{\Gamma}$ generally differs from D by additionally imposed conditions Γ. The resulting index set $\bar{\mathcal{I}} \subseteq \mathcal{I}$ therefore determines the prototypes to be updated and the corresponding prototypes then undergo the learning strategy L resulting in the learning update $\mathcal{L}$.

With the help of this relation we are now able to consider the generic rule (5) in the context of Oja's View: Remember, that we considered the gain g_{ij} to scale and direct the shift $\boldsymbol{s}_{ij}$, while in Oja's Rule ν_{Oja} can be interpreted as to influence the strength of adaptation in (2). Thus, both, gain and ν_{Oja}, aim to regulate the learning behavior. Hence, we require these learning rules to be engineered in such a way, that the prototype adaptation is consistent with the input-output learning relation (8). In consequence, the resulting generic learning rule depends on the decision making $\bar{D}_{\Gamma}$. Suppose, $\mathcal{L}$ is given as $\mathcal{L} = \{\bar{\nu}_{1l}, \ldots, \bar{\nu}_{\kappa l}\}$ with $\bar{\nu}_{il} = \bar{\nu}_{VQ}(\boldsymbol{x}_l)_i$ being valid for the prototype $\boldsymbol{p}_i$ we obtain

$$\Delta \boldsymbol{p}_i = \sigma \cdot \epsilon \cdot \sum_j \bar{\nu}_{ij} \cdot \boldsymbol{s}_{ij} \tag{9}$$

as the appropriate generic description of the learning rule for the considered prototype. This generic ansatz allows respective instances depending on interpretability requirements and/or objective functions to be optimized by the learning process. For the latter case, we could suppose

$$\frac{d\phi_{ij}}{d\boldsymbol{p}_i} = \bar{\nu}_{ij} \cdot \boldsymbol{s}_{ij} \tag{10}$$

to be valid for sufficiently differentiable objective functions O as considered in the Objective View (6), which would fix the σ-value according to the desired gradient scheme (ascent/descent).

Anyway, if (9) does not follow a gradient based update scheme the σ - determination can be integrated in the construction of the gain $\bar{\nu}_{ij}$. For example, in standard LVQ the prototype update does not follow a gradient descent scheme. In consequence, the attraction-repulsing scheme for the best matching class-correct and class-incorrect prototypes can be realized by $\bar{\nu}_{ij}$ being a function depending on class resemblance (see Example 1), thus adjusting the σ - value. The same argumentation as for σ also holds for the learning rate ϵ. For example, attention-based learning in unsupervised VQ [8,10] or cognitive learning strategies for classification learning lead to adjusted learning rates [17]. Consequently, in such cases σ and ϵ have a strong relation to L in (8) and can thus be considered in terms of the gain $\bar{\nu}_{ij}$.

However, if the gradient assumption (10) holds for some unknown/non-specified objective function O, then it can be (re-)constructed by attempting to solute the differential equation

$$\sum_{i}^{\kappa} \mu \cdot \frac{d\phi_{ij}}{d\boldsymbol{p}_i} d\boldsymbol{p}_i = \sum_{i}^{\kappa} \mu \cdot (\bar{\nu}_{ij} \cdot \boldsymbol{s}_{ij})\, d\boldsymbol{p}_i = 0 \tag{11}$$

being of exact type for $\mu = 1$ or inexact for $\mu \notin \{0, 1\}$. Further, from a mathematical point of view, various solutions O of the differential equation may exist depending on the initial condition, i.e. the initial prototype placement influences the final objective implicitly optimized by the algorithm in this case.

Yet, if the objective is available with gradients $\nabla_i O = \frac{d\phi_{ij}}{d\boldsymbol{p}_i}$, the Objective View for the learning input-output relation (8) yields $\mathcal{L} = \{\nabla_1 O, \ldots, \nabla_\kappa O\}$ by applying the mapping $L_{\bar{\mathcal{I}}}(O)$, determining the gradients of O with respect to the selected prototype indices $\bar{\mathcal{I}}$. Further, noting that $\nabla_i O$ is a vector, we can interpret

$$\Delta \boldsymbol{p}_i = \sigma \cdot \epsilon \cdot \nabla_i O \tag{12}$$

for the prototype update. In detail this means, that we are able to interpret the Objective View for VQ also in terms of the underlying input-output relation, which is helpful for creating comprehensible explanations of the learning behavior and, hence, also for the comprehension and trustworthiness of the decision making in the inference (7).

5 Interpreting and Explaining the Generic Learning Rule

So far we combined the Objective and Oja's View by means of the underlying input-output relations for the inference (7) and learning (8) to build the basis for explanations and interpretations of the generic rule (5) and, thus, the resulting decision making. In this regard, we now want to focus on specific aspects of this rule which offer the potential for explanations.

5.1 Gain and Shift

As we already discussed, the gain takes several functionalities in the learning rule, i.e. it scales, regularizes or penalizes. In our view the gain g_{ij} is therefore best understood in terms of a feedback provision. This particularly means, that depending on the conditioned decision making $\bar{D}_\Gamma$, the prototype selection is evaluated through L by various factors like class resemblance or belongingness and the evaluation result $\bar{\nu}_{ij}$, i.e. the networks response in terms of the selection and the external imposed conditions Γ is fed back to the prototype in the learning rule.

Hence, an appropriate definition of the gain and the shift, as well as their interplay, offer a rich explanation potential and give insights into the learning behavior when considered in Oja's View. On the other hand, various VQ models can be constructed in this way, aiming to be *ante-hoc* comprehensible by a careful engineering of the gain and the shift. Yet, it should be emphasized again, that without a respective objective function demanded mathematical properties cannot be guaranteed at all and need to be evaluated additionally.

5.2 The Decision Function

An important detail to explain the learning behavior is the decision function $\bar{D}_\Gamma$ including the imposed conditions Γ and, consequently, a regarding comprehensibility and transparency should be explicitly enhanced. Yet, due to the inherency requirement, this function $\bar{D}_\Gamma$ has a strong impact to the choice of the decision function D in the inference.

In the following we want to consider examples of how to depict a regarding function $\bar{D}_\Gamma$ and its interaction with the subsequent evaluation by the mapping L by considering common examples and we suppose, that $\bar{D}_\Gamma$ selects prototypes based on their belongingness and the externally imposed conditions.

Example 1. For our first example we consider $\bar{D}$ without additional conditions Γ and define $\bar{D}(\mathcal{B}) = \arg\min_{l \in \mathcal{C}} b_l(\boldsymbol{x})$ to be the winner determination where the belongingness $b_i(\boldsymbol{x})$ is chosen as a dissimilarity. Hence, $\bar{D}(\mathcal{B})$ compares the prototypes in terms of their dissimilarity such that $\bar{D}(\mathcal{B})$ is equivalent to a decision tree $T(\mathcal{B})$. As shown in Fig. 1, this yields $T(\mathcal{B}) = s$ with $s = \arg\min_{l \in \mathcal{C}} b_l(\boldsymbol{x})$.

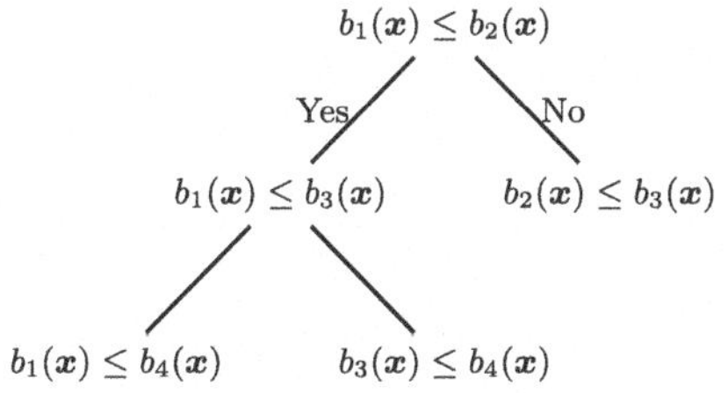

Fig. 1. Schematic explanation how $\bar{D}(\mathcal{B})$ can be transferred into a decision tree $T(\mathcal{B})$ for 4 prototypes

Consequently, depending on the precise definition of $\bar{D}_\Gamma$ we are able to translate this into an easily accessible graph for example-based explanations. Moreover, prototype ordering or rivals could be tracked by understanding (8) as a dynamic system and related back to the result of (7).

A possible extension including Γ is to formulate the conditions as $\Gamma = \{\gamma_1, \gamma_2\}$ with γ_1 = "*class-match*" and γ_2 = "*class-mismatch*" respectively, corresponding to $\bar{D}_\Gamma(\mathcal{B}) = \{T_{\gamma_1}(\mathcal{B}), T_{\gamma_2}(\mathcal{B})\}$. In this extension, $\bar{D}_\Gamma(\mathcal{B})$ yields the indices $\bar{\mathcal{I}} = \{s_1, s_2\}$ of the winners for s_1 a match and s_2 no match of prototype class and class of sample $\boldsymbol{x}$. As a result, the strategy could be adjusted to resemble basic LVQ-1 [15] such that $\mathcal{L} = \{\bar{\nu}_{s_1}(\boldsymbol{x}), \bar{\nu}_{s_2}(\boldsymbol{x})\}$ with $\bar{\nu}_{s_1}(\boldsymbol{x}) = \sigma$ and $\bar{\nu}_{s_2}(\boldsymbol{x}) = -\sigma$ by fixing $\sigma = 1$.

Example 2. This example shows the interaction of $\bar{D}(\mathcal{B})$ and L for unsupervised cases: Here we assume $\bar{D}(\mathcal{B}) = \{T_1(\mathcal{B}), \ldots, T_c(\mathcal{B})\}$ constituting an ensemble of decision trees $T_i(\mathcal{B})$. The tree $T_i(\mathcal{B})$ compares the belongingness $b_i(\boldsymbol{x})$ in terms of a dissimilarity to each remaining belongingness $b_j(\boldsymbol{x})$ as shown in Fig. 2.

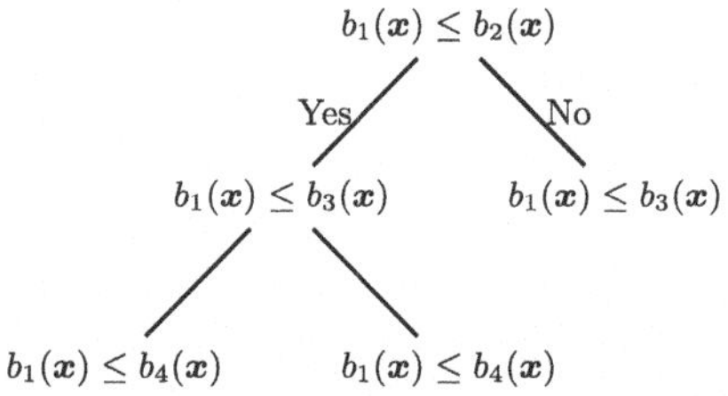

Fig. 2. Schematic decision tree $T_1(\mathcal{B})$ for 4 prototypes

In this case, L then could be defined via $\mathcal{L} = \{h_\lambda(rk(T_1(\mathcal{B}))), \ldots, h_\lambda(rk(T_c(\mathcal{B})))\}$ in which $h_\lambda(rk(T_i(\mathcal{B})))$ is the neighborhood-cooperativeness of the unsupervised neural gas vector quantizer (NG) [18] for prototype $\boldsymbol{p}_i$. Here, the rank-function $rk(T_i(\mathcal{B}))$ simply counts the number of "No" appearing in the tree $T_i(\mathcal{B})$, yielding an tree-based interpretation of NG based on (8). However, the examples show how the interaction of the functions in (8) can be depicted and explained, and, moreover, possibly related to the inference.

5.3 Pitfalls

As discussed before, when learning rules are engineered only based on Oja's View (9), mathematical guarantees might not be given. In this regard, care needs to be taken based on the discussion of *persuasive* and transparent models [7,12] and respective explanations or interpretations. The architecture of a model constructed in such a way may be transparent and intuitive, but possible limitations might not be known or even withhold, due to the lack of an objective function or associated mathematical analysis. In consequence, resulting explanations of the design and decision making may than be used to convince only, while no care was taken with respect to the completeness of the explanation.

6 Summary and Outlook

We presented a generic learning scheme for VQ, which was considered to be derived in an Objective View and with regard to the Oja's View, while showing that both must not be considered separately. Moreover, the resulting scheme is constructed in a way to ensure interpretation and explanation possibilities. In this regard, possible anchors and approaches are given to inspire explanations of existing VQ models (or to think about new models), which can be related to the generic rule (5). This allows the development of comprehensible learning rules with the emphasis on the pitfall of persuasive and complete explanations. Future respective research should focus to provide confidences for the learning related decision making, i.e. to understand such systems in terms of rule-based approaches and evaluate the respective rule-system. In this regard, reject options may be understood in the context of rule evaluation based on approaches like [4,5,9], since these methods directly rely on the decision making.

Acknowledgement. The authors thank for the proofreading of the manuscript by Lynn Reus and Marika Kaden from UAS Mittweida. Further, R.S. acknowledges financial support by the German Ministry of Agriculuture (project - 28DK112B20).

References

1. Artelt, A., et al.: Evaluating robustness of counterfactual explanations. In: 2021 IEEE Symposium Series on Computational Intelligence (SSCI), pp. 01–09 (2021)
2. Biehl, M., Hammer, B., Villmann, T.: Prototype-based models in machine learning. Wiley Interdiscip. Rev. Cogn. Sci. **7**(2), 92–111 (2016)
3. Biehl, M., Ghosh, A., Hammer, B.: Learning vector quantization: the dynamics of winner-takes-all algorithms. Neurocomputing **69**, 660–670 (2006)
4. Brinkrolf, J., Hammer, B.: Interpretable machine learning with reject option. at - Automatisierungstechnik **66**, 283–290 (2018)
5. Duch, W., Setiono, R., Zurada, J.: Computational intelligence methods for rule-based data understanding. Proc. IEEE **92**(5), 771–805 (2004)
6. Erwin, E., Obermayer, K., Schulten, K.: Self-organizing maps: ordering, convergence properties and energy functions. Biol. Cybern. **67**(1), 47–55 (1992)
7. Gilpin, L.H., Bau, D., Yuan, B.Z., Bajwa, A., Specter, M.A., Kagal, L.: Explaining explanations: an overview of interpretability of machine learning. 2018 IEEE 5th International Conference on Data Science and Advanced Analytics (DSAA), pp. 80–89 (2018)
8. Göppert, J., Rosenstiel, W.: Selective attention and self-organizing maps. In: Proceedings of the Neuro-Nimes'94. EC2, Nanterre, France (1994)
9. Hammer, B., Rechtien, A., Strickert, M., Villmann, T.: Vector quantization with rule extraction for mixed domain data (2002)
10. Hasenjäger, M., Ritter, H., Obermayer, K.: Active learning in self-organizing maps. In: Oja, E., Kaski, S. (eds.) Kohonen Maps, pp. 57–70. Elsevier, Amsterdam (1999)
11. Hebb, D.: The Organization of Behavior. A Neuropsychological Theory. John Wiley, New York (1949)
12. Herman, B.: The promise and peril of human evaluation for model interpretability. ArXiv **abs/1711.07414** (2017)

13. Kaden, M., Lange, M., Nebel, D., Riedel, M., Geweniger, T., Villmann, T.: Aspects in classification learning - review of recent developments in learning vector quantization. Found. Comput. Decis. Sci. **39**(2), 79–105 (2014)
14. Karayiannis, N.: An axiomatic approach to soft learning vector quantization and clustering. IEEE Trans. Neural Netw. **10**(5), 1153–1165 (1999)
15. Kohonen, T.: Self-Organizing Maps. Springer, Berlin, Heidelberg (1995). https://doi.org/10.1007/978-3-642-56927-2
16. Lisboa, P., Saralajew, S., Vellido, A., Fernández-Domenech, R., Villmann, T.: The coming of age of interpretable and explainable machine learning models. Neurocomputing **535**, 25–39 (2023)
17. Manome, N., Shinohara, S., Takahashi, T., Chen, Y., Chung, U.I.: Self-incremental learning vector quantization with human cognitive biases. Sci. Rep. **11**(1) (2021)
18. Martinetz, T., Berkovich, S., Schulten, K.: Neural-gas network for vector quantization and its application to time-series prediction. IEEE Trans. Neural Netw. **4**(4), 558–569 (1993)
19. Nova, D., Estévez, P.A.: A review of learning vector quantization classifiers. Neural Comput. Appl. **25**, 511–524 (2013)
20. Oja, E.: Simplified neuron model as a principal component analyzer. J. Math. Biol. **15**, 267–273 (1982)
21. Rudin, C., Chen, C., Chen, Z., Huang, H., Semenova, L., Zhong, C.: Interpretable machine learning: fundamental principles and 10 grand challenges. Stat. Surv. **16**, 1–85 (2022)
22. Ruspini, E.H., Bezdek, J.C., Keller, J.M.: Fuzzy clustering: a historical perspective. IEEE Comput. Intell. Mag. **14**(1), 45–55 (2019)
23. Schneider, P., Biehl, M., Hammer, B.: Adaptive relevance matrices in learning vector quantization. Neural Comput. **21**(12), 3532–3561 (2009)
24. Sokol, K., Vogt, J.E.: (Un)reasonable allure of ante-hoc interpretability for high-stakes domains: transparency is necessary but insufficient for comprehensibility
25. Villmann, T., Haase, S.: Divergence based vector quantization. Neural Comput. **23**(5), 1343–1392 (2011)

Precision and Recall Reject Curves

Lydia Fischer(✉) and Patricia Wollstadt

Honda Research Institute Europe, Carl-Legien-Str. 30,
63073 Offenbach am Main, Germany
lydia.fischer@honda-ri.de

Abstract. For some classification scenarios, it is desirable to use only those classification instances a trained model associates with a high certainty. To evaluate such high-certainty instances, previous work proposed accuracy-reject curves (ARCs). Reject curves evaluate and compare the performance of different certainty measures over a range of thresholds for accepting or rejecting classifications. However, in some scenarios, e. g., data with imbalanced class distributions, the accuracy may not be a suitable evaluation metric and measures like precision or recall may be preferable. We therefore propose reject curves that evaluate precision and recall, the recall-reject curve and the precision-reject curve. Using prototype-based classifiers, we first validate the proposed curves on artificial benchmark data against the ARC as a baseline. We then show on benchmarks and medical, real-world data with class imbalances that the proposed precision- and recall-curves yield more accurate insights into classifier performance than ARCs.

Keywords: reject option · precision and recall · imbalanced data · LVQ

1 Introduction

Machine learning models are used across a wide range of applications, where a common task is to train models for classification of objects. Many of these applications are safety-critical and the reliability and trustworthiness of classifications are particularly important. Such applications come, for example, from the medical domain or high-stakes economic scenarios like logistics for just-in-time production [5]. One method to improve the reliability of classifiers is to calculate a certainty measure for each prediction and use only those predictions with a sufficiently high certainty. The combination of a certainty measure and a threshold that defines the degree of certainty considered sufficient, is called *reject option* [8]. Classifiers with reject options allow to reject data points with a unreliable classification and can improve trust in the application [2,30,34].

When using reject options, classification performance differs for various classifiers, certainty measures, and thresholds. To evaluate reject options, *accuracy reject curves* (ARC)[22] were introduced. ARCs display the accuracy as a function of the rejection rate for a given certainty measure and classifier. They are a

T. Villmann et al. (Eds.): WSOM+2024, LNNS 1087, pp. 163–173, 2024.
https://doi.org/10.1007/978-3-031-67159-3_19

powerful tool for comparing reject options, however, in some scenarios—instead of using the accuracy to judge classifier performance—precision and recall are preferable. This is especially the case for imbalanced data sets, where the evaluation of a reject option using ARCs is inappropriate. We introduce the precision reject curve (PRC) and the recall reject curve (RRC) to close this gap.

In the remainder of this article, we first review related work (Sect. 2) and prototype-based classification, used in the experiments, (Sect. 3). Section 4 introduces the framework of reject options, followed by the introduction of the PRC and RRC for evaluating reject options (Sect. 5). We demonstrate the value of the PRC and RRC in applications to artificial data with available ground-truth distribution, traditional benchmark data, and real-world medical data in Sect. 6, followed by the Conclusion in Sect. 7.

2 Related Work

Chow first introduced reject options with an optimal error-reject trade-off if class probabilities are known [8]. In newer work, classification with rejection is also known as, for example, *selective classification* [11], *abstention* [23,24], and *three way classification* [35]. To evaluate classifiers with reject options, the authors of [22] introduced ARCs, which are widely used today. Alternative approaches to evaluate classifiers with reject options were proposed and investigated in [9,18]. The authors of [18] propose to evaluate reject options by visualising different aspects to find a suitable trade-off between either error rate and rejection rate, cost and rejection rate, or true and false positives (receiver-operator trade-off). For a recent review of reject options, see [19], for a formal treatment, see [15].

Reject options can either be applied as a post-processing step in classification [13,14] or can be integrated into the classifier itself, where the former offers more flexibility as it can be applied to any classifier that allows to define a certainty measure. An instance for classification with integrated rejection is, for example, proposed by [4] and [33], termed *classification by component network* [27].

Using one reject option across the whole input space is called a *global reject option*. A *local reject option* is defined by setting one rejection threshold per class or for different partitions of the input space [14]. The advantage of local reject options is that users can tune thresholds for specific classes or regions, to increase the reliability for important classes or regions. Local reject options and an efficient algorithm for finding optimal local thresholds can be found in [14].

So far we considered reject options for offline, batch-trainable classifiers. For online learning scenarios with drift, there are only first attempts [16] which show that a reject option with a fixed threshold does not increase the performance significantly. Hence more sophisticated methods and measures are needed.

Last, it may be relevant to understand why instances got rejected from classifications. First attempts to use explainability methods in classification with rejection are proposed by [2,3].

3 Prototype-Based Classification

To demonstrate the application of PRC and RRC, we use prototype-based classifiers in the following experiments. We consider this class of models, since it has shown good performance on the considered example data, and well-established certainty measures and global reject options are available [12,13].

Overview of Prototype-based Classifiers: We assume classification tasks in $\mathbb{R}^n$ with Z classes, enumerated as $\{1, \ldots, Z\}$. Prototype-based classifiers are defined as set W of prototypes $(w_j, c(w_j)) \in \mathbb{R}^n \times \{1, \ldots, Z\}$, and $j \in \{1, \ldots, J\}$ that are trained on example data X to represent the data and its class borders.

Every prototype w_j belongs to exactly one class with its class label $c(w_j) \in \{1, \ldots Z\}$. To classify a new data point x, the winner-takes-all-scheme is applied:

$$c(x) = c(w_l) \text{ with } w_l = \arg\min_{w_j \in W} d(w_j, x),$$

where d is a distance measure, often the squared Euclidean distance. Any prototype-based model partitions the feature space into Voronoi cells with one responsible prototype per cell. A data point x falling into a Voronoi cell is assigned the label of the related (closest) prototype, i.e. the winner prototype. The number of prototypes representing a class can be predefined for prototype-based models which leads to a sparse and interpretable representation of the given data X. Heuristics and cost function-based approaches are used as training techniques.

Here, we used extensions of the basic learning vector quantization (LVQ) [26] which uses a heuristic Hebbian learning paradigm. These extensions are the generalized matrix LVQ (GMLVQ) and local generalized matrix LVQ (LGMLVQ), and robust soft LVQ (RSLVQ), which we describe in detail in the following.

GMLVQ and LGMLVQ: By formulating and optimizing explicit cost functions, extensions of LVQ are derived, namely, generalized LVQ (GLVQ) [28] and RSLVQ [31] (next section). For these models convergence guarantees can be given that follow directly from their derivation. For GMLVQ [6], the distance metric is replaced by a general quadratic form, which is also learned during model training. The trained form represents a mapping that puts emphasis on the most discriminative input features and allows to reduce the feature set to the most relevant features only. The LGMLVQ [29] adds a local metric to every prototype and has shown to outperform the GMLVQ in some scenarios.

The GLVQ is based on the formalization as minimization of the cost function

$$E = \sum_i \Phi\left(\frac{d^+(x_i) - d^-(x_i)}{d^+(x_i) + d^-(x_i)}\right), \tag{1}$$

with Φ as a monotonically increasing function, and d^+ and d^- are the distances to the closest prototype, w^+ and w^-, of the correct or incorrect class, for a data point x_i. GLVQ optimizes the location of prototypes by means of a stochastic gradient descent based on the cost function (Eq. 1). For a proof of the learning algorithm's validity at the boundaries of Voronoi cells see [17].

The GMLVQ generalizes the GLVQ to an algorithm with metric adaptation [29]. This generalization takes into account a positive semi-definite matrix Λ in the general quadratic form which replaces the metric d of the GLVQ, i. e. $d(w_j, x) = (x - w_j)^T \Lambda (x - w_j)$. The local version, the LGMLVQ, uses a single metric $d_j(w_j, x) = (x - w_j)^T \Lambda_j (x - w_j)$ for each prototype w_j.

RSLVQ: RSLVQ [31] assumes that data can be modeled via a Gaussian mixture model with labelled types. Based on this assumption, training is performed as an optimization of the data's log-likelihood,

$$E = \sum_i \log p(y_i|x_i, W) = \sum_i \log \frac{p(x_i, y_i|W)}{p(x_i|W)},$$

where $p(x_i|W) = \sum_j p(w_j) \cdot p(x_i|w_j)$ is a mixture of Gaussians with uniform prior probability $p(w_j)$ and Gaussian probability $p(x_i|w_j)$ centered in w_j which is isotropic with fixed variance and equal for all prototypes or, more generally, a general (possibly adaptive) covariance matrix. The probability $p(x_i, y_i|W) = \sum_j \delta_{c(x_i)}^{c(w_j)} p(w_j) \cdot p(x_i|w_j)$ (δ_i^j is the Kronecker delta) describes the probability of a training sample under the current prototype distribution. For a given prediction $\hat{y}$, RSLVQ provides an explicit certainty value $p(\hat{y}|x, W)$, due to the used probability model at the price of a higher computational training complexity.

4 Global Reject Option

A reject option for a classifier is defined by a certainty measure r and a threshold θ, which allows for individual samples to be rejected from classification if the classifier can not make a prediction with a certainty above the threshold. The reject option is further called global if the threshold is constant across the whole input space, i. e., across all classes. (Extending the reject options proposed in the present work to local thresholds is conceivable but beyond the scope of this article (see [14,20]).) Given a certainty measure

$$r : \mathbb{R}^n \to \mathbb{R}, x \longmapsto r(x) \in [0, 1]$$

for a data point x and a threshold $\theta \in \mathbb{R}$, a reject option is defined as a rejection of x from classification iff

$$r(x) < \theta.$$

A rejected data point will not be assigned with a predicted class label. We denote X_θ as all accepted data points with a certainty higher or equal than θ.

In our experiments we use the certainty measures Conf (2) and RelSim (3) that were proposed for prototype-based models in [12]. Additionally, for the artificial data we consider a Bayes classifier that provides ground-truth class probabilities and serves as a baseline (see below).

Conf: Classifiers based on probabilistic models such as RSLVQ provide a direct certainty value of the classification with the estimated probability $\hat{p}(\cdot)$.

$$r_{\text{Conf}}(x) = \max_{1 \leq j \leq Z} \hat{p}(j|x) \in (0,1] \tag{2}$$

RelSim: The relative similarity (RelSim) [12] is based on the GLVQ cost function (1) and considers the distance of the closest prototype (the winner) d^+ and the distance of a closest prototype of any different class d^- for a new unlabelled data point. The winner prototype defines the class label of this new data point, if it is accepted. The measure calculates values according to:

$$r_{\text{RelSim}}(x) = \frac{d^- - d^+}{d^- + d^+} \in [0,1]. \tag{3}$$

Values close to one indicate a certain classification and values near zero point to uncertain class labels. The values of $d^{\pm}$ are already calculated by the used algorithm such that no additional computational costs are caused. Furthermore RelSim (3) depends only on the stored prototypes W and the new unlabelled data point x and no additional storage is needed.

Bayes: The Bayes classifier provides class probabilities for each class provided the data distribution is known. The reject option uses the certainty measure

$$r_{\text{Bayes}}(x) = \max_{1 \leq j \leq Z} p(j|x) \in (0,1] \tag{4}$$

is optimal regarding the error-reject trade-off [8]. In general, the class probabilities are unknown, such that this optimum Bayes reject option can serve as Gold standard for artificially designed settings with a known ground truth, only.

5 Evaluation of Reject Options Using Reject Curves

ARCs [22] are the state of the art for comparing classifiers with a reject option and show the accuracy of a classifier as function of either its acceptance or rejection rate. On the x-axis, ARCs show acceptance rates calculated as $|X_\theta|/|X|$[1], given an applied threshold θ, while on the y-axis, the corresponding accuracy calculated on X_θ is shown. Similarly, the x-axis can show rejection rates $1 - |X_\theta|/|X|$.

ARCs can be easily calculated for binary and multi-class classification scenarios as long as a reject option can be defined for the classifier in question. Formally, the ARC for a given binary data set X is defined as

$$ARC(\theta) : [0,1] \to [0,1], \frac{|X_\theta|}{|X|} \mapsto \frac{TP_\theta + TN_\theta}{|X_\theta|} \tag{5}$$

with $\theta \in \mathbb{R}$, and the true positives (TP_θ) and the true negatives (TN_θ) in X_θ.

[1] X: set of all data; X_θ: set of data with certainty values greater or equal to θ; $|\cdot|$ denotes the cardinality of a set.

While for many classification tasks, in particular for balanced data sets, the accuracy and hence the ARC are suitable techniques, there are scenarios where other evaluation metrics of the classification performance are preferred. For instance, in highly imbalanced scenarios the accuracy of a classifier may be high simply due to—in the worst case—constantly predicting the majority class while the minority class is always misclassified. In such scenarios measures like the $F1$-score, or precision and recall [32] avoid misjudging the performance of a classifier on imbalanced data sets. In [25] a reject curve is proposed with the $F1$-score instead of the accuracy for multi-label settings. Similarly, we introduce the precision reject curve (PRC) and recall reject curve (RRC) as follows,

$$PRC(\theta) : [0,1] \rightarrow [0,1], \frac{|X_\theta|}{|X|} \mapsto \frac{TP_\theta}{TP_\theta + FP_\theta}, \tag{6}$$

$$RRC(\theta) : [0,1] \rightarrow [0,1], \frac{|X_\theta|}{|X|} \mapsto \frac{TP_\theta}{TP_\theta + FN_\theta}. \tag{7}$$

where FP_θ and FN_θ are the false positives and the false negatives in X_θ. Here we show the application of PRCs and RRCs for binary classification only. Similarly to ARCs for multi-class classification (e. g., [13]), both approaches can be extended to multi-class settings, since precision and recall generalize as well [21].

6 Experiments

We demonstrate the usefulness of the PRC and the RRC for the three types of data sets, artificial Gaussian data with known class probabilities, two benchmark data sets, and the Adrenal data set from a real-world medical application.

Gaussian Clusters: The data set contains two artificially-generated, overlapping 2D Gaussian classes, overlaid with uniform noise. Samples are equally distributed over classes. Parameters used were means $\mu_x = (-4, 4.5)$ and $\mu_y = (4, 0.5)$, and standard deviations $\sigma_x = (5.2, 7.1)$ and $\sigma_y = (2.5, 2.1)$.

Tecator data set: The goal for this data set is to predict the fat content (*high* vs. *low*) of a meat sample from its near-infrared absorbance spectrum. Samples are non-equally distributed over classes with 36.0 % vs. 64.0%.

Haberman Data Set: The data set contains 306 instances from two classes (26.5 % vs. 73.5 %) indicating the survival of 5 years and more after breast cancer surgery [10]. Three attributes are present: *age*, *year of operation*, and the *number of positive auxiliary nodes* detected.

Adrenal: The adrenal tumours data set [1] comprises 147 samples composed of 32 steroid marker values as features. The steroid marker values are measured from urine samples using gas chromatography/mass spectrometry. The data comprises two unbalanced classes, namely, patients with benign adrenocortical adenoma (102 or 68.4 % samples) and patients with malignant carcinoma (45 or 30.6 % samples). For medical details we refer to [1,7].

We use a 10-fold repeated cross-validation with ten repeats for our experiments and evaluate models obtained by RSLVQ, GMLVQ, and LGMLVQ with one prototype per class. Since RSLVQ provides probability estimates, we use the certainty measure Conf (2) for rejection. In turn, GMLVQ and LGMLVQ lend itself to the certainty measure RelSim (3).

For the Gaussian data (Fig. 1), the ARC, PRC, and RRC of the RSLVQ model resemble the respective optimal choice obtained from Bayesian classifier closely, for nearly all rejection rates. This is due to little noise and little overlap in the simulated data. For the GMLVQ and the LGMLVQ the shapes of the ARCs, PRCs, and RRCs based on RelSim or Conf are similar to the respective Bayesian baseline results up to a rejection rate of $0.2 = (1 - |X_\theta|/|X|)$, i.e. an acceptance rate of $|X_\theta|/|X| = 0.8$. Performance decreases for lower acceptance rate, which, however, may be less relevant in applications. In sum, our proposed reject curves mirror the optimal performance of the Bayesian classifier closely for acceptance rates that can be considered relevant for practical applications.

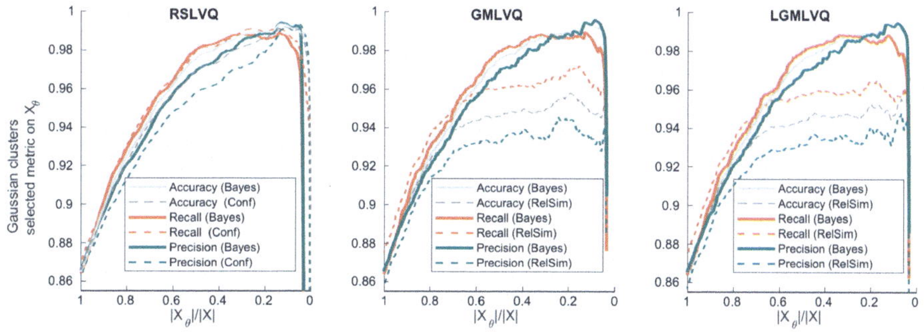

Fig. 1. The averaged reject curves for the different models of the artificial Gaussian data are shown (mean over models in different runs). The solid lines represent the optimal classification performance of a Bayesian classifier. The PRCs and RRCs based on RelSim or Conf perform similar to the optimal ARCs [12,14] for the important regime of at least 80 % accepted data points.

For the benchmark data (Fig. 2), RRCs and PRCs are compared to ARCs [12,14]. Precision is in the same range as accuracy in case of no rejection and for high and medium rejection rates ($|X_\theta|/|X| > 0.3$ for Tecator and $|X_\theta|/|X| > 0.1$ for Haberman). Recall has higher values in case of no rejection and behaves similarly to precision for rejection rates greater than zero. For lower rejection thresholds θ the shapes of the RRCs and the PRCs are monotonically decreasing, with the effect being most prominent for the Tecator data. This behavior can be expected as precision and recall focus on one of the classes instead of both (in binary classification). Here we see that ARC is not a suitable tool to meaningfully evaluate model performance across the full range of θ, as the accuracy overestimates the performance as shown by applying the PRC and RRC.

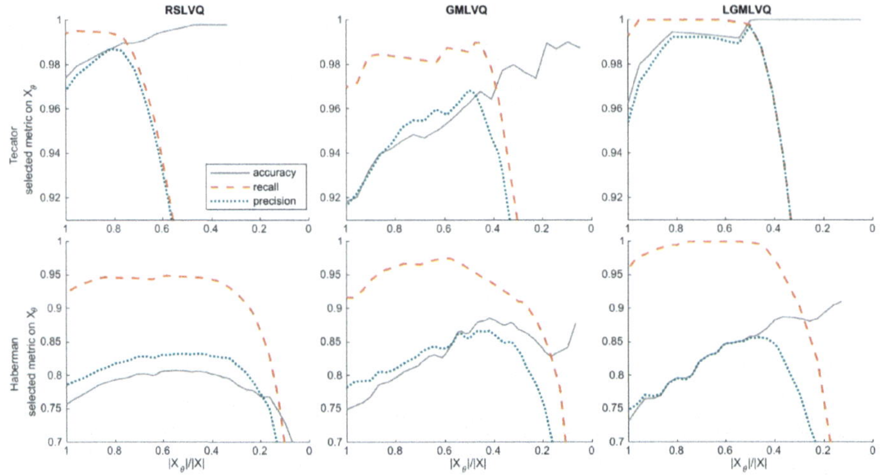

Fig. 2. The image shows results for different LVQ models on benchmark data. The ARCs [12,14] serve as comparison. The PRCs and the RRCs based on RelSim or Conf perform differently for the given set-ups. This reveals interesting insights for the user in order to chose a suited reject threshold for the application scenario at hand.

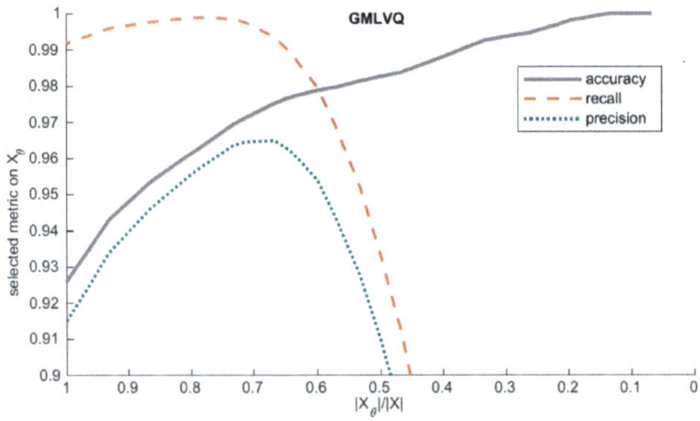

Fig. 3. The averaged curves of the ARC [14] and PRC perform similar in the important regime of at least 80 % accepted data points for the Adrenal data while the RRC has a different shape. The RelSim is used as measure.

Reject curves are relevant for safety-critical scenarios, e. g. in the medical domain. Hence, our last experiment demonstrates the application of the proposed reject curves on a real-world medical data set (Fig. 3). Similar to the benchmark data set, for higher thresholds θ, precision and recall decline while accuracy increases, thus over-estimating performance for low acceptance rates. The experiment shows again that the ARC is not suitable for applications with imbalanced data, while PRC and RRC allow to select meaningful rejection thresholds.

Note that in all experiments, precision and recall develop in a similar fashion. Accordingly, the F1-score will behave in a qualitatively similar fashion as it is the harmonic mean of both quantities. We therefore do not show F1 reject curves here as they will not add insights beyond the PRC and RRC.

7 Conclusion

In this paper we introduced the precision reject curve (PRC) and the recall reject curve (RRC) to evaluate reject options for classification tasks in which precision and recall are the preferred evaluation metrics over the accuracy (e. g., for imbalanced data sets). To demonstrate the suitability of the proposed PRC and RRC, we applied both methods, first, to an artificial data set where we obtained a performance close to ground-truth solutions obtained from Bayesian classifiers. Further, we applied our approach to two popular classification benchmarks, where we showed that our proposed approach allows additional insights into the performance of a classifier on imbalanced data, which could not be obtained from state-of-the-art evaluation using ARCs. Last, we applied the PRC and RRC to one real-world data set from the medical domain where trust in the classification result is particularly important. The latter experiment demonstrates the applicability of our approach as well as its usefulness in a real-world application domain with imbalanced data. Our results show that when evaluating reject options for classification of imbalanced data, the ARC may be misleading, while PRC and the RRC provide meaningful comparisons. Future work may extend the proposed approach to multi-class classification and further evaluation metrics.

Acknowledgements. The authors thank Dylan Fritts and Bradley Brown of HDMA for inspiring discussions. The autors thank Wiebke Arlt for providing the adrenal data set and Stephan Hasler for valuable comments on the manuscript.

References

1. Arlt, W., et al.: Urine steroid metabolomics as a biomarker tool for detecting malignancy in adrenal tumors. J. Clin. Endocrinol. Metab. **96**(12), 3775–3784 (2011)
2. Artelt, A., Brinkrolf, J., Visser, R., Hammer, B.: Explaining reject options of learning vector quantization classifiers. In: IJCCI, pp. 249–261 (2022)
3. Artelt, A., Visser, R., Hammer, B.: Model agnostic local explanations of reject. In: ESANN (2022). https://doi.org/10.14428/ESANN/2022.ES2022-34
4. Bakhtiari, M.M., Villmann, T.: Classification by components including Chow's reject option. In: ICONIP, vol. 1791, pp. 586–596 (2022)
5. Baryannis, G., Validi, S., Dani, S., Antoniou, G.: Supply chain risk management and artificial intelligence: state of the art and future research directions. Int. J. Prod. Res. **57**(7), 2179–2202 (2019)
6. Biehl, M., Ghosh, A., Hammer, B.: Dynamics and generalization ability of LVQ algorithms. J. Mach. Learn. Res. **8**(2), 323–360 (2007)
7. Biehl, M., et al.: Matrix relevance LVQ in steroid metabolomics based classification of adrenal tumors. In: ESANN, pp. 423—428 (2012)

8. Chow, C.: On optimum recognition error and reject tradeoff. IEEE Trans. Inf. Theory **16**(1), 41–46 (1970)
9. Condessa, F., Bioucas-Dias, J., Kovačević, J.: Performance measures for classification systems with rejection. Pattern Recognit. **63**, 437–450 (2017)
10. Dua, D., Graff, C.: UCI machine learning repository (2017). http://archive.ics.uci.edu/ml
11. El-Yaniv, R., Wiener, Y.: On the foundations of noise-free selective classification. J. Mach. Learn. Res. **11**(5), 1605–1641 (2010)
12. Fischer, L., Hammer, B., Wersing, H.: Rejection strategies for learning vector quantization. In: ESANN, pp. 41–46 (2014)
13. Fischer, L., Hammer, B., Wersing, H.: Efficient rejection strategies for prototype-based classification. Neurocomputing **169**, 334–342 (2015)
14. Fischer, L., Hammer, B., Wersing, H.: Optimal local rejection for classifiers. Neurocomputing **214**, 445–457 (2016)
15. Franc, V., Prusa, D., Voracek, V.: Optimal strategies for reject option classifiers. J. Mach. Learn. Res. **24**(11), 1–49 (2023)
16. Göpfert, J.P., Hammer, B., Wersing, H.: Mitigating concept drift via rejection. In: Kůrková, V., Manolopoulos, Y., Hammer, B., Iliadis, L., Maglogiannis, I. (eds.) ICANN 2018. LNCS, vol. 11139, pp. 456–467. Springer, Cham (2018). https://doi.org/10.1007/978-3-030-01418-6_45
17. Hammer, B., Strickert, M., Villmann, T.: Supervised neural gas with general similarity measure. Neural Process. Lett. **21**, 21–44 (2005)
18. Hanczar, B.: Performance visualization spaces for classification with rejection option. Pattern Recognit. **96**, 106984 (2019)
19. Hendrickx, K., Perini, L., der Plas, D.V., Meert, W., Davis, J.: Machine learning with a reject option: a survey. CoRR **abs/2107.11277** (2021)
20. Kummert, J., Paassen, B., Jensen, J., Göpfert, C., Hammer, B.: Local reject option for deterministic multi-class SVM. In: Villa, A.E.P., Masulli, P., Pons Rivero, A.J. (eds.) ICANN 2016. LNCS, vol. 9887, pp. 251–258. Springer, Cham (2016). https://doi.org/10.1007/978-3-319-44781-0_30
21. Manning, C.D., Raghavan, P., Schütze, H.: An Introduction to Information Retrieval. Cambridge University Press, Cambridge, UK (2009)
22. Nadeem, M.S.A., Zucker, J.D., Hanczar, B.: Accuracy-rejection curves (ARCs) for comparing classification methods with a reject option. In: Machine Learning in Systems Biology, pp. 65–81. PMLR (2009)
23. Pazzani, M.J., Murphy, P., Ali, K., Schulenburg, D.: Trading off coverage for accuracy in forecasts: applications to clinical data analysis. In: Proceedings of the AAAI Symposium on Artificial Intelligence in Medicine, pp. 106–110 (1994)
24. Pietraszek, T.: Optimizing abstaining classifiers using ROC analysis. In: Proceedings of the 22nd International Conference on Machine Learning, pp. 665–672 (2005)
25. Pillai, I., Fumera, G., Roli, F.: A classification approach with a reject option for multi-label problems. In: Maino, G., Foresti, G.L. (eds.) ICIAP 2011. LNCS, vol. 6978, pp. 98–107. Springer, Heidelberg (2011). https://doi.org/10.1007/978-3-642-24085-0_11
26. Ritter, H., Kohonen, T.: Self-organizing semantic maps. Biol. Cybern. **61**(4), 241–254 (1989)
27. Saralajew, S., Holdijk, L., Rees, M., Asan, E., Villmann, T.: Classification-by-components: probabilistic modeling of reasoning over a set of components. In: NeurIPS, pp. 2788–2799 (2019)
28. Sato, A., Yamada, K.: Generalized learning vector quantization. Adv. Neural Inf. Process. Syst. **8**, 423–429 (1995)

29. Schneider, P., Biehl, M., Hammer, B.: Adaptive relevance matrices in learning vector quantization. Neural Comput. **21**(12), 3532–3561 (2009)
30. Sendhoff, B., Wersing, H.: Cooperative intelligence-a humane perspective. In: ICHMS, pp. 1–6. IEEE (2020)
31. Seo, S., Obermayer, K.: Soft learning vector quantization. Neural Comput. **15**(7), 1589–1604 (2003)
32. Van Rijsbergen, C.J.: Foundation of evaluation. J. Doc. **30**(4), 365–373 (1974)
33. Villmann, T., et al.: Self-adjusting reject options in prototype based classification. In: Merényi, E., Mendenhall, M.J., O'Driscoll, P. (eds.) Advances in Self-Organizing Maps and Learning Vector Quantization. AISC, vol. 428, pp. 269–279. Springer, Cham (2016). https://doi.org/10.1007/978-3-319-28518-4_24
34. Wang, C., Belardinelli, A., Hasler, S., Stouraitis, T., Tanneberg, D., Gienger, M.: Explainable human-robot training and cooperation with augmented reality. In: CHI Conference on Human Factors in Computing Systems, pp. 1–5 (2023)
35. Yao, Y.: Three-way decision: an interpretation of rules in rough set theory. In: Wen, P., Li, Y., Polkowski, L., Yao, Y., Tsumoto, S., Wang, G. (eds.) RSKT 2009. LNCS (LNAI), vol. 5589, pp. 642–649. Springer, Heidelberg (2009). https://doi.org/10.1007/978-3-642-02962-2_81

K Minimum Enclosing Balls for Outlier Detection

Daniel Staps[1(✉)], Thomas Villmann[1], and Benjamin Paaßen[2,3]

[1] Saxon Institute for Computational Intelligence and Machine Learning, University of Applied Sciences, Mittweida, Germany
{staps,villmann}@hs-mittweida.de

[2] Faculty of Technology, Bielefeld University, Bielefeld, Germany
bpaassen@techfak.uni-bielefeld.de

[3] Educational Technology Lab, German Research Center for Artificial Intelligence, Kaiserslautern, Germany

Abstract. Outlier detection means to characterize the distribution of inliers exactly enough such that outliers stand out. A natural and interpretable model is an enclosing ball that includes non-outliers but excludes outliers. However, a single enclosing ball is insufficient for data with multiple, distinct clusters or modes. Therefore, we propose a model with K enclosing balls, each of which can cover a separate cluster of data. We investigate two optimization schemes, one based on stochastic gradient descent and one based on alternating optimization and quadratic programming. On synthetic and real-world data we compare our proposed K minimum enclosing ball (K-MEB) approach to baselines from the literature.

Keywords: Outlier Detection · One-Class Classification · Minimum Enclosing Ball · Support Vector Description · Interpretability

1 Introduction

Outlier detection is concerned with detecting rare events that deviate sufficiently from normal operation to warrant special attention, such as fraud in banking systems, intrusions in cybersecurity, or illnesses in medicine [21]. In principle, outlier detection is a binary classification problem of distinguishing outliers (negative class) from "normal" samples (positive class). However, our training data will only include very few (if any) examples of the outlier class which do not cover all possible outliers. Instead, we wish to characterize the distribution of the positive class as precisely as possible to identify outliers by contrast to the positive class [21]. To improve the characterization of the positive class, deep model have emerged which achieved ever higher detection accuracy, such as deep autoencoder approaches [24] or deep support vector description [12].

However, detection accuracy is not the only concern in outlier detection. In many practical scenarios, detecting an outlier implies serious practical consequences, such as legal persecution in case of bank fraud and cyber security, or

T. Villmann et al. (Eds.): WSOM+2024, LNNS 1087, pp. 174–184, 2024.
https://doi.org/10.1007/978-3-031-67159-3_20

medical interventions in medicine. Accordingly, we ought to be able to justify, explain, and interpret an outlier detection decision [15,23]. Therefore, we aim for an outlier detection model which achieves suitable detection accuracy while remaining interpretable.

Our starting point is the classic minimum enclosing ball (MEB) problem [17]: We try to find the smallest ball that encloses (almost) all positive samples while excluding as many outliers as possible. In line with learning vector quantization models, we can interpret the center of the ball w as a prototype for the positive class, as it lies right in the center of the represented data [4,18]. Further, we can easily measure the margin from the decision boundary by comparing the distance $d(x, w)$ of a point x to the center w with the ball's radius r. If $d(x, w) > r$, we consider x an outlier. Note that the margin to the decision boundary is also crucial for adversarial robustness [3,13,20].

Unfortunately, it is easy to find examples where an MEB underfits the data. In particular, if the positive class consists of multiple clusters or modes, the ball must either focus on a single cluster or include a lot of volume that does not include positive data. For example, in Fig. 1, we can only cover the blue points with one ball if we also include outliers.

One approach to improve model fit is kernelization [17,21]. In the feature space of a radial basis function kernel, for example, we can solve the problem in Fig. 1 with a single ball. However, this loses some interpretability [4]. What does the center of our ball look like? Where is the class boundary? Such questions are much easier to answer in the primal space. Therefore, we propose a different approach, namely multiple balls. This enables us to fit a distinct ball to each cluster or mode of the positive distribution (refer to Fig. 1).

More precisely, our contribution is as follows.

- We formalize the K minimum enclosing ball (K-MEB) problem as a prototype model for outlier detection.
- We introduce two optimization approaches for the K-MEB problem, namely a stochastic gradient approach (SGD) and an alternating optimization with quadratic programming.
- We evaluate the proposed K-MEB approach against real data baselines.

2 Related Work

The MEB problem has already been widely used for outlier detection, most prominently as the basis for support vector data description (SVDD) [17], which finds a hypersphere (in primal or dual space) that contains all positive data points. We use the same formalization and the same dual formulation as in SVDD, but we generalize it to K balls. A generalization of SVDD, the mixture of support vector data description (mSVDD) is proposed in [8]. Here, they also take a mixture of hyperspheres for novelty detection, introducing conditional probabilities for a probabilistic approach to make decisions. Yet, we rely on a geometric interpretation and quadratic programming.

Our approach also bears resemblance to the set cover problem of [1]. In this problem, they try to cover a set of points with a minimum amount of ϵ balls without including any points of the negative class. Similarly, in [9] the set cover problem was tackled for classification tasks based on a cost function introduced in [22] for metric learning. The kernel density estimation (KDE) [10,11] estimates the underlying density by an ensemble of kernel functions with fixed centers at the data points positions. Our formulation is different in that we fix the number of balls K as a hyperparameter but let the radius of each ball be flexible and allow for arbitrary locations of the centers in the data space.

An approach which takes flexible balls into account too, is the supervised vector quantization one class classifier (SVQ-OCC) [16]. When optimizing their parameters by stochastic gradient descent learning, the center of the balls is learned by the data density explicitly and the radius of each ball is adapted by local underlying probability distributions. However, we explicitly minimize the radius of each ball, while the centers of the balls arise implicitly.

3 Method

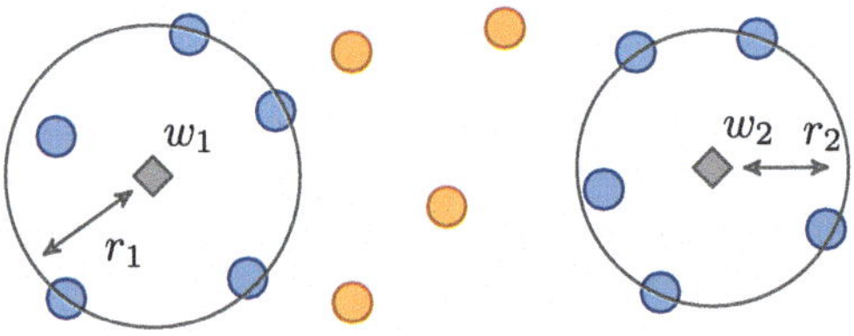

Fig. 1. An illustration of the K-minimum enclosing ball problem with 2 balls. Blue points include positive samples, orange points negative samples. Diamonds illustrate the center of balls.

Our goal is to find K minimum enclosing balls that cover the positive samples in our dataset but do not cover outliers. More precisely, let $x_1, \ldots, x_m \in \mathbb{R}^n$ be our data points and $y_1, \ldots, y_m \in \{-1, +1\}$ be a label where $y_i = -1$ if x_i is an outlier and $y_i = +1$, otherwise. Further, let $w_1, \ldots, w_K \in \mathbb{R}^n$ be the centers of our K balls and $r_1, \ldots, r_K \in \mathbb{R}^+$ be the corresponding radii. We formalize our K-minimum enclosing ball problem as follows.

$$\begin{aligned} \min_{r_1,\ldots,r_K,w_1,\ldots,w_K} \quad & \sum_{k=1}^{K} r_k^2 \\ \text{s.t.} \quad & \forall i \text{ where } y_i = +1 : \exists k : d(x_i, w_k)^2 \leq r_k^2 \\ & \forall i \text{ where } y_i = -1 : \forall k : d(x_i, w_k)^2 \geq r_k^2, \end{aligned} \tag{1}$$

where d is the Euclidean distance on $\mathbb{R}^n$. In other words: We try to minimize the radii r_k of all balls while ensuring that every positive sample is included in at least one ball and every negative sample is included in no ball (Fig. 1).

To make this formulation more compact, we introduce auxiliary variables $h_{k,i} \in \{-1, 0, 1\}$ where $h_{k,i} = -1$ if $y_i = -1$ and $h_{k,i} = +1$ if $y_i = +1$ and point x_i should be covered by ball k. Further, we replace our strict side constraints with soft constraints and punish violations with the hinge loss. Overall, we obtain the following, relaxed problem:

$$\min_{r_1,\dots,r_K,w_1,\dots,w_K,\boldsymbol{H}} \quad \sum_{k=1}^{K} r_k^2 + \sum_{i=1}^{m} C_i \cdot \left[h_{k,i} \cdot \left(d(x_i, w_k)^2 - r_k^2\right)\right]_+ \tag{2}$$

$$\text{s.t.} \quad \forall i \text{ where } y_i = +1 : \sum_{k=1}^{K} h_{k,i} \geq 1 \land \forall k : h_{k,i} \in \{0, 1\}$$

$$\forall i \text{ where } y_i = -1 : \forall k : h_{k,i} = -1,$$

where $[x]_+ = \max\{0, x\}$ denotes the hinge loss and $C_i > 0$ are hyperparameters which weigh the importance of each soft constraint. Note that the first side constraint ensures that every positive sample is included in the receptive field of at least one ball, similar to the set cover formalization of [1].

Note that problem (1) is hard to solve directly due to the integer variables $h_{k,i}$. Therefore, we need heuristic approaches. As a theoretical basis for our heuristics, we establish a lemma regarding the optimal setting of $h_{k,i}$ for constant balls.

Lemma 1. *For constant $r_1, \dots, r_K$ and $w_1, \dots, w_K$, and assuming that for all i the $\arg\min_l \left(d(x_i, w_l)^2 - r_l^2\right)$ is unique, the following setting of $\boldsymbol{H}$ is globally optimal.*

$$h_{k,i} = \begin{cases} -1 & \text{if } y_i = -1 \\ +1 & \text{if } y_i = +1 \land k = \arg\min_l \left(d(x_i, w_l)^2 - r_l^2\right) \\ 0 & \text{otherwise} \end{cases} \tag{3}$$

Proof. First, we define the auxiliary functions $f_i(\tilde{\boldsymbol{H}}) = \sum_{k=1}^{K} C_i \cdot \left[h_{k,i} \cdot \left(d(x_i, w_k)^2 - r_k^2\right)\right]_+$ *for all i. Next, note that for any i with $y_i = +1$, we have exactly one k_i, namely $k_i = \arg\min_l d(x_i, w_l)^2 - r_l^2$, with $h_{k_i,i} = 1$. For all other $k \neq k_i$, we have $h_{k,i} = 0$. Therefore, we obtain $f_i(\boldsymbol{H}) = C_i \cdot \left[d(x_i, w_{k_i})^2 - r_{k_i}^2\right]_+$.*

Now, assume that our claim is wrong, meaning that there exists some $\boldsymbol{H}'$ that is feasible according to (2) and has lower objective function value. Note that the objective function is $\sum_{k=1}^{K} r_k^2 + \sum_{i=1}^{m} f_i(\boldsymbol{H}')$. Because all r_k are constant, there must exist at least one i, such that $f_i(\boldsymbol{H}') < f_i(\boldsymbol{H})$. If $y_i = -1$, we obtain $h'_{k,i} = h_{k,i} = -1$ for all k due to the side constraints of (2). Therefore, $f_i(\boldsymbol{H}') = f_i(\boldsymbol{H})$, which is a contradiction. If $y_i = +1$, we note that there must exist at least one k with $h'_{k,i} = 1$, due the first side constraint of (2). Further, note that all terms $C_i\left[d(x_i, w_l)^2 - r_l^2\right]_+$ are non-negative because $C_i > 0$ and the hinge loss is non-negative. Accordingly, we obtain:

$$f_i(\boldsymbol{H}') \geq C_i \cdot \left[d(x_i, w_k)^2 - r_k^2\right]_+ \geq C_i \cdot \left[\min_l d(x_i, w_l)^2 - r_l^2\right]_+ = f_i(\boldsymbol{H})$$

This is a contradiction to $f_i(\boldsymbol{H}') < f_i(\boldsymbol{H})$. Since no other options for y_i remain, the objective function value for $\boldsymbol{H}'$ must be at least as large as for $\boldsymbol{H}$, which implies that $\boldsymbol{H}$ is a global optimum, as claimed.

Using this information about $\boldsymbol{H}$, we introduce two optimization approaches, a stochastic gradient descent, and an alternating optimization.

3.1 Stochastic Gradient Descent

Our first approach is a stochastic gradient optimization inspired by Generalized Learning Vector Quantization (GLVQ) [14]. In GLVQ, we consider the data points one by one and pull the closest prototype from the same class closer while pushing the closest prototype from a different class away, both of which corresponds to a stochastic gradient descent step on the GLVQ loss function [14]. Accordingly and similarly to a GLVQ approach for one-class classification proposed in [16], here we also consider points x_i one by one (randomly sampled). We set $h_{k,i}$ according to (3) for all k.

Then, we take a stochastic gradient descent step based on the objective function (2) for point i. This corresponds to moving all balls away from negative samples and the closest ball towards a positive sample—similar to GLVQ. In practice, we do not use vanilla stochastic gradient descent but, rather, a scheme like Adam [6].

3.2 Alternating Optimization

Our second approach is an alternating optimization scheme, similar to K-Means [5] or hypersphere mixtures [8]. First, we initialize $w_1, \ldots, w_K$ randomly in the data space and set $r_1 = \ldots = r_K = 1$. Then, we set $\boldsymbol{H}$ according to (3), which corresponds to the expectation step of [8].

Next, we keep $\boldsymbol{H}$ fixed and only optimize $w_1, \ldots, w_K$ and $r_1, \ldots, r_K$. Because the side constraints only concern $\boldsymbol{H}$, this optimization decouples into K independent minimum enclosing ball problems. For every k, we need to solve

$$\begin{aligned} \min_{r_k, w_k, \vec{\epsilon}} \quad & r_k^2 + \sum_{i=1}^{m} C_i \cdot \epsilon_i && (4) \\ \text{s.t.} \quad & \epsilon_i \geq h_{k,i} \cdot \left(d(x_i, w_k)^2 - r_k^2\right) \wedge \epsilon_i \geq 0 && \forall i, \end{aligned}$$

where ϵ_i is a slack variable that represents the hinge loss. This is a quadratically constrained, linear problem and, hence, convex. Therefore, we can find a global optimum with standard tools. This yields Algorithm 1. It is easy to show that this algorithm converges.

Theorem 1. *Algorithm 1 is guaranteed to stop and to stop at a local optimum of* (2).

Proof. Thanks of Lemma 1*, we know that line 9 of Algorithm* 1 *never increases the objective function of* (2)*. Further, because problem* (4) *is convex, line 14–16 yield a global optimum for* k*. Since this holds for all* k*, and since the sum of the objective functions of* (4) *corresponds to the objective function of* (2)*, lines 13–17 can not increase the objective function, either. Therefore, the main loop in lines 8–18 generates a non-increasing sequence of objective function values. Finally, if the condition in line 10 applies, no local change in* $\boldsymbol{H}$ *can improve the objective function anymore. Further, no local change in* w_k *or* r_k *can improve the objective function—otherwise, it would have been performed in the previous iteration. Therefore, if Algorithm* 1 *stops, it stops at a local optimum. Further, the algorithm must stop because there are only finitely many configurations of* $\boldsymbol{H}$ *and we can never repeat a* $\boldsymbol{H}$*. If we would repeat an* $\boldsymbol{H}$*, either line 11 would be triggered, or* w_k *and* r_k *would have changed to increase the objective function—which can not happen due to the previous argument.*

3.3 Kernel Approach

While we advocate for learning in the primal space for interpretability, we also provide a pure kernel version of our approach for completeness' sake. To do so, in line with [17], we construct the Wolfe dual of problem (4), which yields

$$\begin{aligned} \min_{\vec{\alpha}_k} \quad & \vec{\alpha}_k^T \cdot \mathrm{diag}(\vec{h}_k) \cdot \boldsymbol{K} \cdot \mathrm{diag}(\vec{h}_k) \cdot \vec{\alpha}_k - \mathrm{diag}(\boldsymbol{K})^T \cdot \mathrm{diag}(\vec{h}_k) \cdot \vec{\alpha}_k \\ \text{s.t.} \quad & \vec{h}_k^T \cdot \vec{\alpha}_k = 1 \wedge \vec{C} \geq \vec{\alpha}_k \geq 0, \end{aligned} \tag{5}$$

where $\alpha_{k,i}$ is the Lagrange multiplier for point i of the kth minimum enclosing ball problem, $\mathrm{diag}(\vec{h}_k)$ is a $m \times m$ diagonal matrix with entries $h_{k,1}, \ldots, h_{k,m}$, $\boldsymbol{K}$ is the kernel matrix, and $\mathrm{diag}(\boldsymbol{K})$ is the diagonal of $\boldsymbol{K}$.

Note that, for a linear kernel, we recover the centers of each ball via the Karush-Kuhn-Tucker condition $w_k = \sum_{i=1}^{m} \alpha_{k,i} \cdot h_{k,i} \cdot x_i$. To recover the radii r_k, we can solve (2) with respect to r_k only.

For inference, we re-write the squared distance $d(w_k, x_i)^2$ based on the kernel $\boldsymbol{K}$ as follows:

$$\begin{aligned} d(w_k, x_i)^2 =& d\Big(\sum_{j=1}^{m} \alpha_{k,j} \cdot h_{k,j} \cdot x_j, x_i\Big)^2 \\ =& \vec{\alpha}_k^T \cdot \mathrm{diag}(\vec{h}_k) \cdot \boldsymbol{K} \cdot \mathrm{diag}(\vec{h}_k) \cdot \vec{\alpha}_k - 2 \cdot \sum_{j=1}^{m} \alpha_{k,j} \cdot h_{k,j} \cdot k_{j,i} + k_{i,i}. \end{aligned} \tag{6}$$

Algorithm 1. Our alternating optimization to solve problem (2).

```
1: function K-MEB(data vectors x_1, ..., x_m, labels y_1, ..., y_m)
2:     for k ← 1, ..., K do
3:         Select random i with y_i = +1.
4:         Set w_k ← x_i; r_k ← 1.
5:     end for
6:     while True do
7:         Set H according to (3).
8:         if H remained unchanged in line 8 then
9:             return w_1, ..., w_K, r_1, ..., r_K.
10:        end if
11:        for k ← 1, ..., K do
12:            Solve Problem (4) for ball k to obtain w_k and r_k.
13:        end for
14:    end while
15: end function
```

3.4 Prediction

To predict the label for any new data points x, we simply need to compute the distance $d(x, w_k)^2$ to all centers w_k and check whether $d(x, w_k)^2 \leq r_k^2$ for any ball k, that is, whether x lies inside any of the K balls. If so, we predict $y = +1$. Otherwise, we predict $y = -1$. For the kernel version, we use (6) to compute the relevant distances.

4 Experiments and Discussion

For numerical experiments, we selected semantically meaningful data sets for outlier evaluation described in [2], which are selected from the UCI repository[1].

Table 1. Data sets with semantic outliers after preprocessing.

data set	class vs. outlier	—class data—	—outlier data—	—attributes—
Annthyroid	'Normal' vs. two types of hypothyroidism	(train) 3488 (test) 3178	284 250	21 21
Arrhythmia	Healthy vs. 12 types of cardiac arrhythmia	243	199	277
Cardiotocography	'Normal' vs. suspect, pathological	1654	471	21
HeartDisease	Healthy vs. heart problems	160	137	13
Hepatitis	Survival vs. fatal	105	24	17
PageBlocks	Text vs. non-text	4913	560	10
Parkinson	Healthy vs. Parkinson	48	147	22
SpamBase	Non-Spam vs. Spam	2788	1813	57
Wilt	Land covers vs. Diseased Trees	(train) 4265 (test) 313	74 187	5 5

[1] https://archive.ics.uci.edu/.

We applied usual preprocessing techniques to handle missing values. The details of the data setting can be inferred from Table 1. Furthermore, we applied a z-score normalization for the rectified datasets. In comparison to [2], which focuses on the evaluation of unsupervised outlier detection, we also exploit some outliers and compare to supervised outlier detection algorithms. In particular we compare: KDE [10,11], SVDD [17], mSVDD [8], SVQ-OCC [16] and our proposed K-MEB algorithm.

We analyze the models' abilities with different kernels: *Gaussian kernel* - KDE; *linear kernel* - SVDD, mSVDD, K-MEB, *rbf-kernel* - SVDD, K-MEB and the *Euclidean distance* - K-MEB, SVQ-OCC. Further, K-MEB is once optimized by a quadratic solver and, additionally, by stochastic gradient descent learning (SGD). To implement the kernel approach for SGD and a given kernel κ, we rewrite the distances as $d_\kappa^2(\mathbf{x}, \mathbf{y}) = \kappa(\mathbf{x}, \mathbf{x}) - 2 \cdot \kappa(\mathbf{x}, \mathbf{y}) + \kappa(\mathbf{y}, \mathbf{y})$ similar to [19].

For each setting, the best hyperparameters are determined by a grid search. If train and test sets are given by the UCI repository, they are used (as it can be seen in Table 1). Otherwise, a 5-fold cross validation is applied.

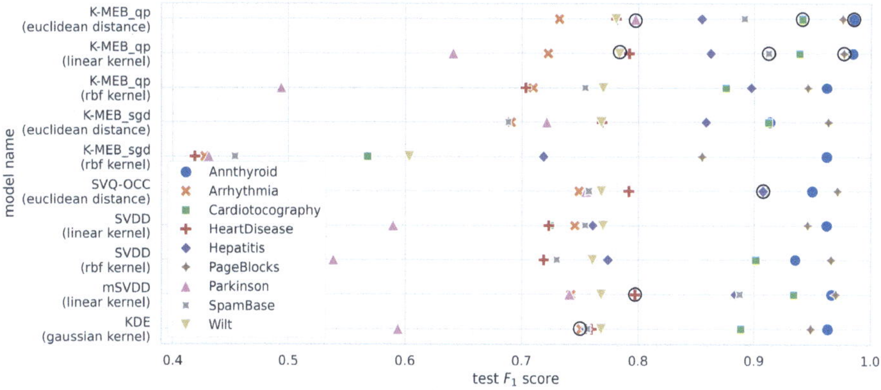

Fig. 2. Visualization of the experiment results. The horizontal axis shows the F_1 score for the test set of the best trained models. Each row refers to a model (qp: quadratic problem, sgd: stochastic gradient descend). Results are shown for nine data sets per model setting. The best result for each data set is surrounded by a black circle.

In Fig. 2 the F_1 score test results for each model and the nine data sets are depicted. The proposed K-MEB learned by the alternating optimization scheme with either the *Euclidean distance* or the *linear kernel* yield similar results, meaning the F_1 score is roughly similar for every data set, which demonstrates the consistency of both versions of our model. Moreover, both approaches achieve the best results for six out of the nine data sets, which are Annthyroid, Cardiotocography, PageBlocks, Parkinson, SpamBase and Wilt. Hepatitis is best solved by the SVQ-OCC algorithm, HeartDisease by the mSVDD and Arrhythmia by the KDE model. In general, the other algorithms show mostly similar or nearly as good results. An exception is the Parkinson data set, where in comparison to KDE and SVDD, our K-MEB with the *rbf-kernel* shows bad results. Especially the SGD version may have encountered numerical instabilities with the *rbf-kernel* during learning, which can be observed in Fig. 2 as the F_1 score is worse for a lot of data sets. Further, we observe that K-MEB optimized by alternating optimization shows overall slightly better results for different kernels. Although K-MEB with *rbf-kernel* is better then the SGD version, altogether the kernels with possibility for geometrical interpretation are better. This is reflected by the SVQ-OCC trained with the *Euclidean distance*, SVDD and mSVDD with the *linear kernel*, too.

Figure 3 exemplarily shows the classification results for the K-MEB algorithm using the *Euclidean distance* learned by alternating optimization on the Parkinson data set [7]. For visualization purpose, the model is only trained with two features from the data set: MDVP:Fo(Hz) - Average vocal fundamental frequency and MDVP:Jitter(%) - Variation in fundamental frequency, which are features responsible for voice disorder detection. The hyperspheres envelope the positive class (healthy), while the negative class (parkinson) is forced to remain outside. The hyperspheres centers are representing the data points of the positive class, which can be seen the best at the top left two prototypes. A counterexample is the right most prototype. Yet, the problem may occur, that multiple separated data points are all assigned to this one prototype: During learning, the algorithm tries to include all the positives and cannot sacrifice only one data point to envelope the nearby data points on the top left.

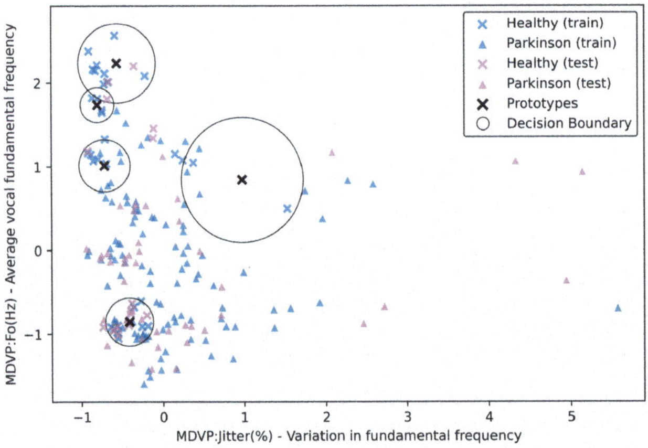

Fig. 3. Illustration of the classification results for the K-MEB (*Euclidean distance*, quadratic solver) on Parkinson data [7]. For explanations – see text.

5 Conclusion

We introduced different optimization approaches for the K-MEB problem. We conclude that the approach based on alternating optimization with quadratic programming equipped with the *Euclidean distance* and the *linear kernel* are best suited to solve the stated outlier detection problems. Additionally, both variants are the most stable to learn and unfold the best possibilities to geometrically interpret the model parameters. Future work is to cope with hyperspheres, which got stuck on data points such that the prototypes lose its interpretability. Furthermore, questions need to be answered on: How to choose the number of k balls? How to best interpret the balls in more than two dimensions?

Acknowledgements. This work was funded by the German Aerospace Research Center (projects AIMS/DAIMLER grant: 50WK2270A and AIMS/IAI-XPRESS grant: 50WK2270D and AID grant: 50WM2354) and the German Federal Agency for Agriculture and Food (project KIOSens grant: 28DK112B20). Additional thanks to Lynn Reuss, Marika Kaden and Thomas Davies for providing comments on styling and writing of the paper and Ronny Schubert for supporting discussions.

References

1. Bien, J., Tibshirani, R.: Prototype selection for interpretable classification. Ann. Appl. Stat. **5**(4), 2403–2424 (2011)
2. Campos, G.O., et al.: On the evaluation of unsupervised outlier detection: measures, datasets, and an empirical study. Data Min. Knowl. Disc. **30**, 891–927 (2016)
3. Göpfert, C., Göpfert, J.P., Hammer, B.: Adversarial robustness curves. In: Cellier, P., Driessens, K. (eds.) ECML PKDD 2019. CCIS, vol. 1167, pp. 172–179. Springer, Cham (2020). https://doi.org/10.1007/978-3-030-43823-4_15
4. Hofmann, D., Schleif, F.M., Paaßen, B., Hammer, B.: Learning interpretable kernelized prototype-based models. Neurocomputing **141**, 84–96 (2014)
5. Jain, A.K.: Data clustering: 50 years beyond k-means. Pattern Recogn. Lett. **31**(8), 651–666 (2010)
6. Kingma, D.P., Ba, J.: Adam: a method for stochastic optimization. arXiv arxiv:1412.6980 (2014)
7. Little, M.: Parkinsons. UCI Machine Learning Repository (2008)
8. Nguyen, D., Lai, V., Nguyen, K., Le, T.: Mixture of hyperspheres for novelty detection. Viet. J. Comput. Sci. **3**(4), 223–233 (2016)
9. Paaßen, B., Villmann, T.: Prototype selection based on set covering and large margins. Mach. Learn. Rep. **14**(MLR-03-2021), 35–42 (2021)
10. Parzen, E.: On estimation of a probability density function and mode. Ann. Math. Stat. **33**(3), 1065–1076 (1962)
11. Rosenblatt, M.: Remarks on some nonparametric estimates of a density function. Ann. Math. Stat. 832–837 (1956)
12. Ruff, L., et al.: Deep one-class classification. In: Dy, J., Krause, A. (eds.) Proceedings of the 35th International Conference on Machine Learning (ICML), pp. 4393–4402 (2018)
13. Saralajew, S., Holdijk, L., Villmann, T.: Fast adversarial robustness certification of nearest prototype classifiers for arbitrary seminorms. In: Larochelle, H., Ranzato, M., Hadsell, R., Balcan, M., Lin, H. (eds.) Proceedings of the 34th Conference on Neural Information Processing Systems (NeurIPS 2020), vol. 33, pp. 13635–13650. Curran Associates, Inc. (2020)

14. Sato, A., Yamada, K.: Generalized learning vector quantization. In: Touretzky, D., Mozer, M., Hasselmo, M. (eds.) Proceedings of the 8th International Conference on Advances in Neural Information Processing Systems (NIPS), vol. 8, pp. 423–429. MIT Press (1995)
15. Shekhar, S., Shah, N., Akoglu, L.: Fairod: fairness-aware outlier detection. In: Proceedings of the 2021 AAAI/ACM Conference on AI, Ethics, and Society, pp. 210–220 (2021)
16. Staps, D., Schubert, R., Kaden, M., Lampe, A., Hermann, W., Villmann, T.: Prototype-based one-class-classification learning using local representations. In: Proceedings of the IEEE International Joint Conference on Neural Networks (IJCNN) - Padua. IEEE Press (2022)
17. Tax, D.M., Duin, R.P.: Support vector data description. Mach. Learn. **54**(1), 45–66 (2004)
18. Villmann, T., Bohnsack, A., Kaden, M.: Can learning vector quantization be an alternative to svm and deep learning? - recent trends and advanced variants of learning vector quantization for classification learning. J. Artif. Intell. Soft Comput. Res. **7**(1), 65–81 (2016)
19. Villmann, T., Haase, S., Kaden, M.: Kernelized vector quantization in gradient-descent learning. Neurocomputing **147**, 83–95 (2015)
20. Voráček, V., Hein, M.: Provably adversarially robust nearest prototype classifiers. In: Chaudhuri, K., Jegelka, S., Song, L., Szepesvari, C., Niu, G., Sabato, S. (eds.) Proceedings of the 39th International Conference on Machine Learning, pp. 22361–22383 (2022)
21. Wang, H., Bah, M.J., Hammad, M.: Progress in outlier detection techniques: a survey. IEEE Access **7**, 107964–108000 (2019)
22. Weinberger, K.Q., Saul, L.K.: Distance metric learning for large margin nearest neighbor classification. J. Mach. Learn. Res. **10**(9), 207–244 (2009)
23. Zimek, A., Filzmoser, P.: There and back again: outlier detection between statistical reasoning and data mining algorithms. WIREs Data Min. Knowl. Disc. **8**(6) (2018)
24. Zong, B., et al.: Deep autoencoding gaussian mixture model for unsupervised anomaly detection. In: International Conference on Learning Representations (2018)

Probabilistic Models with Invariance

Alexander Gepperth(✉)

Applied Computer Science Department, Fulda University of Applied Sciences, Leipzigerstr.123, 36037 Fulda, Germany
alexander.gepperth@cs.hs-fulda.de
https://www.gepperth.net/alexander

Abstract. This proof-of-concept work generalizes the concept of invariance, as used in contrastive learning, to fully probabilistic models (such as, e.g., mixture models) that explicitly describe data distributions in an interpretable fashion, and whose main applications are density estimation (e.g., outlier detection), sampling and tractable inference. Invariance allows allows probabilistic models to operate at a lower effective model complexity, and therefore to deal with more complex (image) data. In this article, we propose iGMM, a Gaussian Mixture Model (GMM) that explicitly incorporates invariance into its loss, which is a generalization of the conventional GMM log-likelihood. When constructing hierarchies of conventional GMM and iGMM instances, we obtain invariance properties that are reminiscent of simple and complex cells in the mammalian visual cortex. We show, by experiments on the MNIST and FashionMNIST dataset, that GMM-iGMM hierarchies can faithfully sample from learned data distributions even if the iGMM is invariant to some aspects of the data, and demonstrate that outlier detection performance is strongly enhanced in GMM-iGMM hierarchies.

1 Introduction

Recent years have seen tremendous advances in what is now called *generative models*, whose primary function after training is to generate (i.e., sample) images that are similar to training images. Most prominent representatives of generative models are doubtlessly GANs and VAEs.

In contrast to this, we term models that aim at directly capturing the data density $p(\boldsymbol{x})$ *fully probabilistic models.* Important examples are mixture models such as Gaussian Mixture Models (GMMs). Probabilistic models offer efficient sampling, outlier detection as well as tractable inference, although model complexity needs to be significantly higher for realistic data (e.g., images) to produce meaningful results[10].

1.1 Motivation: Complex Image Data for CL

There are several ways to motivate interest in more powerful probabilistic models. Besides conceptual reasons, probabilistic models are of particular interest

T. Villmann et al. (Eds.): WSOM+2024, LNNS 1087, pp. 185–195, 2024.
https://doi.org/10.1007/978-3-031-67159-3_21

in the field of continual learning (CL, [13]), i.e., machine learning from non-stationary data distributions. An obvious concern is, e.g., the detection of non-stationarity onsets via outlier detection. Furthermore, efficient sampling is relevant for replay methods in CL where there is a dedicated "generator" for remembering "past" samples instead of storing them. Lastly, inference is required when querying for samples that are similar to the ones that are currently processed. Probabilistic model such as sum-product networks (SPNs, see [1,8,15]) are unfortunately not yet able to process complex datasets that contain significant structural variations such as color shifts or translations. DCGMMs have had limited success in handling SVHN [2], although the color variations in particular pose significant challenges.

Probabilistic models have a drawback: they aim to model the full data distribution, including aspects that are completely irrelevant to downstream tasks. For example, in a digit recognition problem, the polarity or foreground/background color of digits is irrelevant, yet, e.g., a GMM will spend considerable resource capturing these aspects. If probabilistic models such as GMMs could be trained to ignore certain aspects of the input distribution (e.g., shifts, background color, rotation, ...), this kind of *invariance* would free up significant resources for modeling the task-relevant aspects of a data distribution. In a digit recognition task, such GMMs with invariance (iGMMs) would mainly model aspects related to shape. Please see Fig. 1(left) for a visualization of the concept.

1.2 Related Work

To the best of our knowledge, no works that integrate invariance into probabilistic models have been proposed. In the following text, we shall discuss models that cover some aspects of probabilistic models, which we understand to include sampling, outlier detection and probabilistic inference:

VAEs and GANs. GANS and VAEs do not achieve their considerable successes by drawing samples from an explicit learned representation $p(\boldsymbol{x})$ of the data density. VAEs in particular rather model the prior probability for latent variables $p(\boldsymbol{z})$, the conditional decoder density $p(\boldsymbol{x}|\boldsymbol{z})$ and the conditional encoder density $p(\boldsymbol{z}|\boldsymbol{x})$. While VAEs cannot perform tractable inference, they can use importance sampling to approximate $p(\boldsymbol{x})$, although at considerable computational cost. GANs cannot perform density estimation, outlier detection or tractable inference since they are not formulated in terms of probabilities. A key issue with GANs and VAEs is that they can use the standard tools of DNNs, such as max-pooling layers, to ensure invariance at least to positional shifts or rotations.

Flow-Based Models. Flow-based models [4,6,9] try to transform a known distribution into the observed data distribution by means of adaptive invertible transformations, usually mediated by special deep neural networks. Indeed, competitive sampling has been demonstrated for complex image data, whereas outlier detection suffers from problems that are not fully understood [17]. Since the overall distribution is still intractable, inference seems to difficult as well. As the overall transformation mediated by any flow-based model is supposed to be invertible, standard DNN mechanisms like max-pooling layers are not applicable for ensuring invariance.

Sum-Product Networks. Sum-product networks (sometimes termed probabilistic circuits or Einsum networks) [1,8,15] are directed trees whose leaves represent tractable probability densities, and whose structure aims at efficiently expressing hierarchical mixture models. They are thus a true probabilistic model and can perform all associated functions. However, they are lacking in expressiveness when modeling real-world images: even for MNIST and similar datasets, generated samples do not seem overly realistic, while more complex problems like SVHN are currently out of scope. We know of no works that attempt to incorporate invariance into SPNs or related models.

Mixture Models and DCGMMs. The most well-known mixture models are Gaussian Mixture Models (GMMs). They implement a "flat" hierarchy with just a single layer of computation that is global in the sense that the whole image is processed at once. There have been several attempts to create hierarchical GMMs, see., e.g., [11,12,14], which have however only been applied to synthetic data. Mixture of Factor Analyzers (MFAs) are an extension of GMMs that has been applied to model the CelebA face dataset [10], although using extensive pretraining. To the best of our knowledge, the only attempt to create a hierarchical *and* convolutional GMM is that of [2], with application to the SVHN dataset.

1.3 GMMs with Invariance (iGMMs)

As a key concept, we propose an invariance-enforcing extension to vanilla GMMs that we term iGMM. iGMM instances are trained with augmented data and encourage the best-matching GMM component to be identical for all augmented versions of the same sample. Since iGMMs are based on Mahalanobis distances, the best-matching component will automatically get more *dissimilar* to other samples by consequence. Instead of applying iGMMs to image content directly, it is advantageous to use them as layers in a hierarchy together with a vanilla GMM layer, see Fig. 1.

1.4 Contributions

This article is a proof-of-concept work operating on simple datasets. The conceptual contribution is a proposal for an invariance-enforcing generalization of the GMM log-likelihood that can be used for SGD training. In addition, we describe a simple way to sample from a trained GMM-iGMM hierarchy. Experimental contributions include a training procedure based on GMM-iGMM hierarchies for the efficient learning of invariances, as well as an analysis of invariance, sampling and outlier detection properties of such hierarchies.

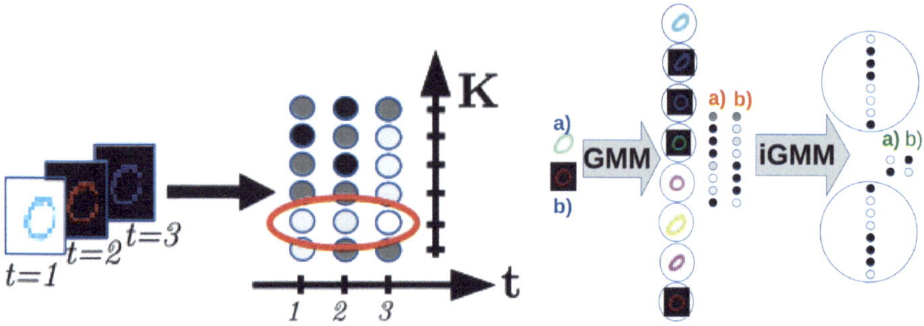

Fig. 1. **Left**: visualization of iGMM invariance, shown for $A = 3$ augmented versions of a single sample passed through an iGMM with $K = 6$ components, resulting in $AK = 18$ component activities (log-probabilities) color-coded as filled circles. Invariance to augmentations translates into the requirement that one component (red ellipse) should have consistently higher activity for all augmented versions of a sample. The iGMM loss is formulated such that the component with maximal average (taken over $t = 1, 2, 3$) activity is adapted. **Right:** response to different stimuli (blue letters)) in a GMM-iGMM cascade. A GMM with 6 components (centroids shown inside large circles) is directly trained on color-augmented MNIST digits, followed by an iGMM with two components (centroids again in large circles). GMM responses (marked by red letters, strength indicated by brightness) show selectivity to slant, color and polarity. iGMM responses (indicated by green letters) on the other hand just react to different polarities but are invariant to other image properties. (Color figure online)

2 Methods

2.1 Data Augmentation in General

We assume the existence of several parameterized transformations $\tau_i(\boldsymbol{p})$ such that a training example $\boldsymbol{x}_\nu$ can be augmented as $T_\mathcal{V}(\boldsymbol{x}_\nu) \equiv \tau_1(\boldsymbol{v}_1) \circ \tau_2(\boldsymbol{v}_2) \circ \ldots (\boldsymbol{x}_\nu), \boldsymbol{v}_i \in \mathcal{V}$ according to the set of parameter vectors $\mathcal{V}$. By creating a fixed number A of augmented samples for every $\boldsymbol{x}_\nu$ in a mini-batch of N samples, we create multi-viewed mini-batches of AN samples for training. The indices of all augmented version of sample $\boldsymbol{x}_\nu$ are given by $P(\nu)$, with $|P(\nu)| = A$. Conversely, the source index of an augmented sample $\boldsymbol{x}_n$ is given by $\nu(n)$. The set of parameter vectors is chosen randomly and uniformly from the ranges defined for a particular transformation.

2.2 Transformations for Data Augmentation

Channel swap either exchanges two random RGB channel in the whole image or does nothing, with each of these four possibilities chosen with equal probability. The parameter vector is either $\boldsymbol{v} = (a, b)^T$ for a channel swap, or $\boldsymbol{v} = (-1, -1)^T$ for no swap.

Brightness inversion is a deterministic transformation which simply performs the operation $x \to 255 - x$ on all channels simultaneously. The parameter vector is a single number indicating whether or not to perform this transformation.

Random image shifts are performed with shifts $\Delta_x, \Delta_y \in [-3, 3]$ using cyclic boundary conditions. The parameter vector is simply $\boldsymbol{v} = (\Delta_x, \Delta_y)^T$.

2.3 Data and Preprocessing

All experiments are based on the MNIST and FashionMNIST datasets.

MNIST [7] consists of 60 000 28×28 gray scale images of handwritten digits (0–9). We promote all images to RGB by repeating the single channel 3 times, and multiplying channels by normalized random factors in the $[0, 1]$ interval. Prior to normalization, one randomly chosen factor is multiplied by a random number between 3 and 6 so as to induce stronger colors.

Fashion-MNIST [16] consists of 60.000 images of clothes in 10 categories and is structured like MNIST.

Both datasets are transformed into colored versions of themselves by setting all non-black pixels to a randomly selected shade of red, green or blue. For each sample, we create $A = 5$ augmented versions using channel swap, image shift and brightness inversion operations.

2.4 GMMS with Invariance: iGMMs

Vanilla GMMs try to represent the target distribution as a weighted sum of multi-variate normal distributions with centroids $\boldsymbol{\mu}$ and covariance matrices $\boldsymbol{\Sigma}$: $p(\boldsymbol{x}) = \sum_k \pi_k \mathcal{N}_k(\boldsymbol{x}; \boldsymbol{\mu}_k, \boldsymbol{\Sigma}_k)$. A trained GMM outputs the posterior probability $\gamma_k(\boldsymbol{x}_i) = \frac{\pi_k \mathcal{N}(\boldsymbol{x}; \boldsymbol{\mu}_k, \boldsymbol{\Sigma}_k)}{\sum_j \pi_j \mathcal{N}(\boldsymbol{x}; \boldsymbol{\mu}_j, \boldsymbol{\Sigma}_j)}$ for each image $\boldsymbol{x}_i$ and GMM component k.

Generalizing the conventional GMM log-likelihood loss, the iGMM loss pushes the best-matching component k^* to be identical for all augmented versions of the same sample $\boldsymbol{x}_\nu$, and to be different for other samples as much as possible:

$$\mathcal{L} = \sum_n \log \sum_k (\pi_k \mathcal{N}_k(\boldsymbol{x}_n)) \approx \sum_n \log \max_k (\pi_k \mathcal{N}_k(\boldsymbol{x}_n)) = \sum_n \max_k \log (\pi_k \mathcal{N}_k(\boldsymbol{x}_n)) \quad (1)$$

$$\to \sum_\nu \max_k \sum_{n \in A(\nu)} \log(\pi_k \mathcal{N}_k(\boldsymbol{x}_n)). \quad (2)$$

The first expression of Eq. (1) represents the incomplete-data loss for a vanilla GMM. Subsequently, we employ the max-component approximation of the sum as outlined in [2], which removes exponentials and brings the max operation in front of the log. In the second row, we generalize to the case of multiple augmented versions of a single sample $\boldsymbol{x}_\nu$. The mean over all augmented versions $\boldsymbol{x}_n$ of the sample $\boldsymbol{x}_\nu$ will have high values if the same component consistently shows high log-probabilities. This is similar to constrastive learning [5], except that we do not need to enforce dissimilarity to other samples due to the distance-based nature of the loss Eq. (1).

An iGMM can be applied to image data directly, in which case components tend to averages over augmented samples. This leads to blurred or unrecognizable sampling results. A better ways of using iGMMs is to apply them to sparse data, such as posterior probabilities computed by a vanilla GMM. This allows to construct a GMM-iGMM sampling chain of hierarchical priors as outlined in Sect. 2.5 on the one hand, and on the other hand ensures that learned iGMM components remain sparse as well, which facilitates interpretability and sampling, please see also Fig. 2.

2.5 GMM-IGMM Hierarchies

GMM-iGMM hierarchies are trained using a separate loss for the GMM and iGMM layer. In contrast to standard procedure for, e.g., DNNs, both losses are not added to form a single overall loss, but optimized in complete independence. Otherwise, adaptation of GMM parameters would be impacted by the iGMM loss as well, which has proven to prevent convergence. Both "layers" are linked through the GMM posterior probabilities which form the input to the iGMM. Posterior probabilities or responsibilities are computed as $\gamma_k(\boldsymbol{x}) = \frac{p_k(\boldsymbol{x})}{\sum_j p_j(\boldsymbol{x})}$ and are therefore normalized and bounded in the $[0, 1]$ range. Both GMM and iGMM are trained by SGD according to [3], and iGMM adaptation is prevented until the GMM annealing parameter $\sigma(t)$ has reached a value of 0.8, which means that a certain convergence has already been achieved.

Sampling in a GMM-iGMM hierarchy is performed by generating a sample in the iGMM layer, which is used as a prior over component selection probabilities when sampling from the GMM layer. When sampling from a certain GMM component directly, this hierarchical prior mechanism is disabled.

3 Experiments

We conduct the experiments with GMMs, iGMMs and GMM-iGMM hierarchies according to the following principles: most importantly, all optimization is carried out by SGD (not Expectation-Maximization or similar schemes). Image data are converted to vectorial form by simple flattening (e.g., a $28 \times 28 \times 3$ images becomes a 2352-dimensional vector). When image augmentation as described in Sect. 2.1 is just performed for training, whereas only unaugmented colored samples are used during evaluation. In hierarchies, the forward pass for one samples (regardless of training or testing) consist always of passing the vector of GMM responsibilities as input to the iGMM. Furthermore, the GMM and iGMM loss are optimized in parallel, and adaptation is inhibited for the iGMM until the GMM has converged sufficiently.

3.1 Preliminary Experiment: Naive Application of iGMMs

For this experiment, we directly train an iGMM instance on MNIST data augmented as described in Sect. 2.3. To illustrate the developed component selectivities, we sample from certain components of this iGMM and display the resulting

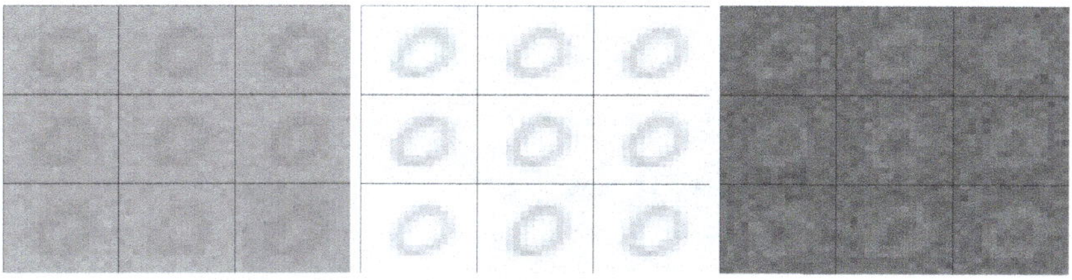

Fig. 2. Component selectivities of an iGMM instance that has been directly trained on augmented data, visualized by sampling from these components.

images. See Fig. 2. We can observe that components tend to get selective to "average samples", which are mainly blank since polarity-inversed samples will cancel each other out in the average.

3.2 Formation of Simple/complex Cells in GMM-IGMM Hierarchies

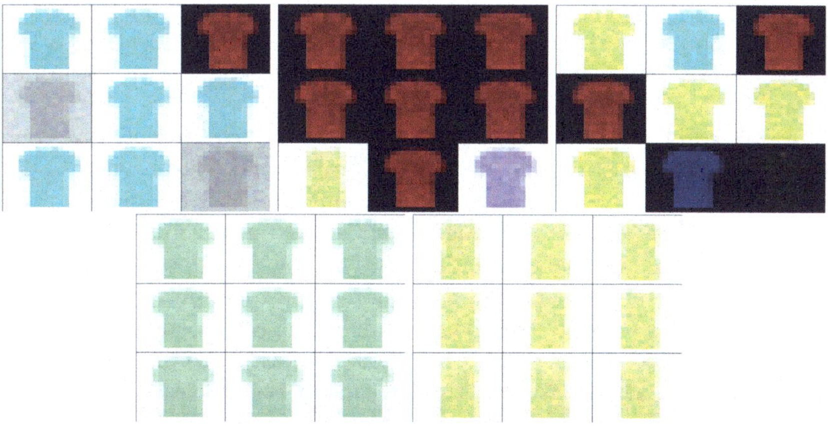

Fig. 3. Component selectivities in a GMM-iGMM hierarchy visualized by sampling from different iGMM (upper row) and GMM (lower row) components after training on colored FashionMNIST. We can observe invariance, to different degrees, for iGMM components but none for GMM components.

Motivated by the negative outcome of Sect. 3.1, we perform a similar experiment with a GMM-iGMM hierarchy as described in Sect. 2.5. Again, we determine component selectivities through selectively sampling from those components, both for the GMM and the iGMM "layer". Sampling from the GMM layer is performed in the standard GMM way, whereas sampling from the iGMM layer proceeds in two steps: first of all, the iGMM is queried for a sample, again in the

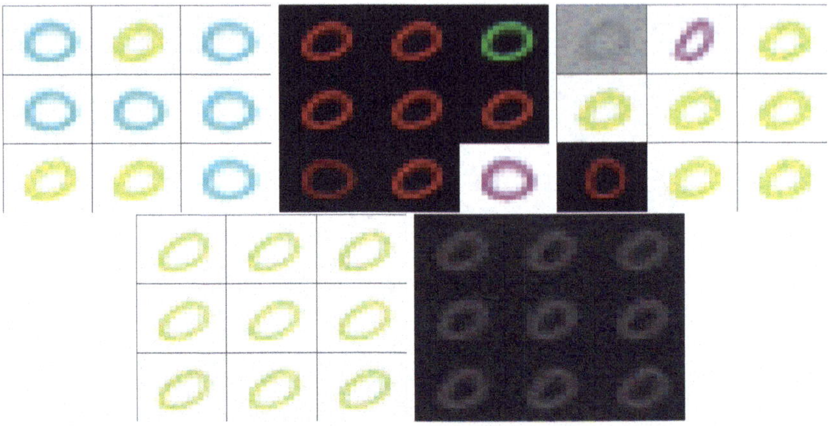

Fig. 4. Component selectivities in a GMM-iGMM hierarchy visualized by sampling from different iGMM (upper row) and GMM (lower row) components after training on colored MNIST. We can observe invariance, to different degrees, for iGMM components but none for GMM components.

standard way. This sample is then taken to represent a multinomial distribution over GMM components, from which we draw a sample that identifies a GMM component. This component will produce the final sampling result from the multivariate Gaussian distribution it represents.

In Fig. 3 and Fig. 4, we observe very interesting behavior: iGMM components are selective to stimuli of multiple polarities and colors, whereas GMM components are selective just for one given configuration w.r.t. these axes. It is furthermore discernible that, although iGMM components do have preferences for, e.g., certain colors or polarities, they always show some degree of invariance. This is reminiscent of the behavior of simple/complex cells in the visual cortex: whereas simple cells respond only to test stimuli of a given polarity, complex cells respond to the same stimuli but with inversed polarity. We find it noteworthy that such behavior arises naturally through unsupervised learning.

3.3 Explicitly Measuring Invariance

In this section, we introduce am explicit measure of invariance for assessing the differences between GMM and iGMM "layer" in a GMM-iGMM hierarchy. The measure is recorded over the course of training and evaluated on the test set as well. The invariance measure we propose essentially measures the maximal number of times any component is best-matching unit (BMU) for all augmented version of a given sample, averaged over all distinct samples in the test set. We define the invariance measure based on the posterior probabilities $\gamma(\boldsymbol{x})$:

$$\chi_k(\boldsymbol{x}) = \begin{cases} 1 & k = \arg\max_j \gamma_k(\boldsymbol{x}) \\ 0 & \text{else} \end{cases} \tag{3}$$

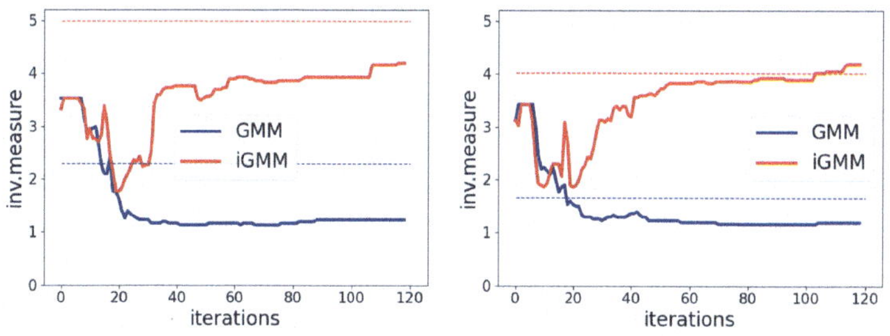

Fig. 5. Invariance measure C for the GMM and iGMM layer plotted over the course of 120 training epochs for MNIST(left) and FashionMNIST(right). Invariance measures plotted on the respective test sets after epoch 120 are shown as dashed lines. We observe that the iGMM layer always shows higher invariance.

Using the notation form Sect. 2.1, we can count the number of times any component was BMU for all augmented version of a single sample as $c_k(\nu) = \sum_{x_n \in P(\nu)} \chi_k(x_{n(\nu)})$. The desired invariance measure C is then found as: $C = \mathbb{E}_\nu \max_k c_k(\nu)$. When training a GMM-iGMM hierarchy on MNIST and FashionMNIST, we obtain the curves depicted in Fig. 5, from which we can observe a markedly higher invariance for the iGMM layer.

4 Discussion

From the experiments of the previous section, we conclude that GMM-iGMM hierarchies are capable of capturing simple invariances, and to nevertheless perform faithful sampling in the input domain. The modification of the iGMM loss to reward invariance appears to be beneficial for describing abstract concept with fewer components, as shown by the success of the outlier detection experiment. Some specific topics seem worth highlighting here:

Choice of Datasets. Since this is a proof.of-concept work, we do not perform experiment on complex real-world datasets such as CIFAR since these are hard to model even for dedicated generators such as VAEs. When moving to deep convolutional hierarchies as outlined in [2], such datasets will within the range of capabilities of our probabilistic models.

Assessment of Selectivities Through Sampling. Since probabilistic models always have the capacity to sample, we consider this an interesting and natural tool for visualizing component selectivities. This is in stark contrast to deep neural networks, where one has to resort to complex procedure like deep dreaming to analyze selectivities of specific neurons.

Link to Constrastive Supervised Learning. The approach to training iGMMs is similar to the one taken by constrastive learning approaches. Certainly, data augmentation is a common property, but also the explicit modeling of invariances. We believe that by generalizing GMM-iGMM hierarchies to deeper convolutional hierarchies, we can model invariances at a local image level and learn feature compared to those obtained by contrastive learning.

Reproducibility. We will release TF2-based Python code to reproduce all experiments.

References

1. Butz, C.J., Oliveira, J.S., dos Santos, A.E., Teixeira, A.L.: Deep convolutional sum-product networks. In: Proceedings of the AAAI Conference on Artificial Intelligence, vol. 33, pp. 3248–3255 (2019)
2. Gepperth, A.: A new perspective on probabilistic image modeling. In: International Joint Conference on Neural Networks(IJCNN) (2022)
3. Gepperth, A., Pfülb, B.: Gradient-based training of gaussian mixture models in high-dimensional spaces (2020)
4. Grathwohl, W., Chen, R.T., Bettencourt, J., Sutskever, I., Duvenaud, D.: Ffjord: free-form continuous dynamics for scalable reversible generative models. arXiv preprint arXiv:1810.01367 (2018)
5. Khosla, P., et al.: Supervised contrastive learning. Adv. Neural. Inf. Process. Syst. **33**, 18661–18673 (2020)
6. Kingma, D.P., Dhariwal, P.: Glow: generative flow with invertible 1×1 convolutions. Adv. Neural Inf. Process. Syst. **31** (2018)
7. LeCun, Y., Bottou, L., Bengio, Y., Haffner, P.: Gradient-based learning applied to document recognition. Proc. IEEE **86**(11), 2278–2324 (1998)
8. Peharz, R., et al.: Einsum networks: fast and scalable learning of tractable probabilistic circuits. In: International Conference on Machine Learning, pp. 7563–7574. PMLR (2020)
9. Rezende, D., Mohamed, S.: Variational inference with normalizing flows. In: International Conference on Machine Learning, pp. 1530–1538. PMLR (2015)
10. Richardson, E., Weiss, Y.: On GANs and GMMs. Adv. Neural Inf. Process. Syst. **31**, 5847–5858 (2018)
11. Tang, Y., Salakhutdinov, R., Hinton, G.: Deep mixtures of factor analysers. In: Proceedings of the 29th International Conference on Machine Learning, ICML 2012, vol. 1, pp. 505–512 (2012)
12. Van Den Oord, A., Schrauwen, B.: Factoring variations in natural images with deep Gaussian mixture models. Adv. Neural. Inf. Process. Syst. **4**(January), 3518–3526 (2014)
13. Van de Ven, G.M., Tolias, A.S.: Three scenarios for continual learning. arXiv preprint arXiv:1904.07734 (2019)
14. Viroli, C., McLachlan, G.J.: Deep Gaussian mixture models. Stat. Comput. **29**(1), 43–51 (2019). https://doi.org/10.1007/s11222-017-9793-z
15. Wolfshaar, J., Pronobis, A.: Deep generalized convolutional sum-product networks. In: International Conference on Probabilistic Graphical Models, pp. 533–544. PMLR (2020)

16. Xiao, H., Rasul, K., Vollgraf, R.: Fashion-MNIST: a novel image dataset for benchmarking machine learning algorithms, pp. 1–6 (2017). http://arxiv.org/abs/1708.07747
17. Zhang, L., Goldstein, M., Ranganath, R.: Understanding failures in out-of-distribution detection with deep generative models. In: International Conference on Machine Learning, pp. 12427–12436. PMLR (2021)

Optimizing YOLOv5 for Green AI: A Study on Model Pruning and Lightweight Networks

Bangguo Xu[1,2], Simei Yan[1,2], Liang Liu[1,2], and Frank-Michael Schleif[1,2(✉)]

[1] Department of Computer Science and Business Information Systems, UAS Würzburg-Schweinfurt, Würzburg, Germany
frank-michael.schleif@fhws.de

[2] Center for Artificial Intelligence and Robotics, UAS Würzburg-Schweinfurt, Würzburg, Germany

Abstract. To achieve state-of-the-art performance, deep learning models are becoming increasingly complex, leading to a significant increase in demand for high-performance computing resources and, in turn, concerns about environmental impact. In this context, the concept of Green AI has been proposed, which advocates optimizing performance by improving model efficiency, rather than relying solely on an increase in computing resources, thereby reducing the impact on the environment. Object detection is a research hotspot in the field of computer vision. This paper focuses on the commonly used YOLOv5 network in object detection, optimizing the YOLOv5 model through pruning and the use of lightweight networks. Implemented on a campus image dataset, a balance between reducing computational load and maintaining accuracy was achieved. The experimental results confirm that strategic model pruning and thoughtful network architecture selection can produce environmentally responsible and computationally efficient deep learning models without significantly reducing performance, aligning with the goals of GreenAI. All related code of the project is available at: https://github.com/xbgthws/Green-AI-project.git.

Keywords: Transfer Learning · MobileNet · ShuffleNet · GhostNet

1 Introduction

In recent years, the complexity and performance of Artificial Intelligence (AI) and machine learning models have continually increased, leading to a significant demand for high-performance computing resources. This demand encompasses not only a reliance on high-performance computers but also involves substantial electricity consumption and corresponding cooling requirements, thereby having a considerable environmental impact. This issue becomes particularly pronounced in training large-scale deep learning models, where energy consumption and carbon emissions have become focal points of concern. The concept of "Green

T. Villmann et al. (Eds.): WSOM+2024, LNNS 1087, pp. 196–205, 2024.
https://doi.org/10.1007/978-3-031-67159-3_22

AI [19]," first introduced by Roy Schwartz et al. in 2019, emphasizes the importance of minimizing the environmental impact while advancing and innovating in AI technology. Green AI advocates for optimizing model performance through innovation and efficiency improvements rather than solely relying on increasing computational resources.

In this context, YOLOv5 (You Only Look Once version 5), was proposed as a popular object detection algorithm [14]. This algorithm is widely recognized for its exceptional speed and accuracy, making it particularly suitable for real-time object detection tasks. We aim to explore how to reduce energy consumption and enhance the efficiency of YOLOv5 through algorithmic and network structure optimizations, without significantly sacrificing performance, which aligns with the objectives of Green AI.

2 Related Work

This section briefly reviews some works related to our study, the LabelImg system, YOLO network, other detectors and transfer learning.

2.1 LabelImg

[1] LabelImg, an open-source tool developed in Python, provides an intuitive interface for image annotation, essential for computer vision tasks. In our dataset, we annotated all classes within each image, saving these annotations in the YOLO format, which records the bounding box coordinates and corresponding class labels. LabelImg has been widely adopted by recent research literature to perform object detection tasks related to YOLO networks, e.g. [2,9,21].

2.2 YOLO

In the domain of deep learning for object detection, there are generally two approaches: two-stage and one-stage methods. The two-stage methods, such as R-CNN, Fast R-CNN, and Faster R-CNN, first generate region proposals and then classify them, generally achieving high accuracy but at a slower pace. On the other hand, one-stage methods like YOLO and SSD (Single Shot MultiBox Detector) detect objects directly in a single step, suitable for application with high real-time requirements.

The YOLO series, initiated by Joseph Redmon et al. in 2016 with YOLOv1 [14], innovatively solves object detection as a regression problem, obtaining the position, class, and confidence probability of objects in an image through a single inference. YOLOv2, or YOLO9000 [15], introduced in 2017, improved in accuracy, speed, and detectable object categories. YOLOv3 [16], released in

[1] LabelImg is currently no longer in active development and will not receive any new feature updates or fixes. Its legacy and core functionalities have been integrated into Label Studio, which offers a broader range of data annotation capabilities.

2018, enhanced the network architecture with multi-scale detection. YOLOv4 [1], developed in 2020 by Alexey Bochkovskiy et al., integrated various previous research techniques for a balance of speed and accuracy. YOLOv5, by Glenn Jocher and others, is an open-source project from Ultralytics with versions ranging from n to x, scaling up in parameter size and performance.

2.3 Pretrained Models and Fine-Tuning

Transfer learning is a method that allows knowledge trained on one task to be applied to another related task [24]. Pretrained models are a commonly used tool in transfer learning, which refer to models that have been trained on large datasets. By using a pre-trained model as a starting point and further training (i.e., fine-tuning) on a task-specific dataset, you can speed up the learning process and improve the model's performance on new tasks.

Recent studies have shown that initializing with transferred features can improve the generalization performance of a model, even after substantial fine-tuning on a new task [22]. This has been identified as a generally useful technique for enhancing the performance of deep neural networks. Additionally, convolutional neural networks are capable of learning rich mid-level image features that are transferrable to a variety of visual recognition tasks [13]. In the application of YOLOv5, this means that the model does not need to learn all features from scratch but can quickly adapt to new tasks by fine-tuning a pretrained network, thus identifying specific objects in new datasets, reducing training time, and improving learning efficiency. In scenarios where labeled data is scarce, transfer learning offers a significant advantage to researchers or projects lacking the resources for extensive data annotation, thereby saving computational resources and lowering barriers to entry [17]. In this way, YOLOv5 can effectively utilize transfer learning to achieve higher performance and efficiency in new domains or tasks.

3 Preliminary

This paper aims to explore the possibility of realizing Green AI by optimizing the YOLOv5 network, especially to improve model efficiency while minimizing the impact on accuracy. In the previous section, we have discussed the relevant tools used in this experiment. Next, we discuss in depth the theoretical foundations and practical applications of model pruning and lightweight neural networks. These strategies are crucial for optimizing the YOLOv5 network to adapt to resource-constrained environments.

3.1 Pruning

In our research, we focus on adjusting the importance coefficients of specific layers in neural networks. These coefficients determine the contribution of each feature to the network's output. By identifying and gradually reducing the importance of less contributive features [12], we can optimize the network's structure.

By integrating L1 regularization into the loss function as an additional penalty, we encourage the model to recognize and reduce the dependency on unimportant features [4]. This process, guiding the sparsity of the gamma parameters in the batch-normalization layers, further refines the network's architecture. As a result, unimportant features are removed from the final network structure, simplifying the model. Applying this technique to the YOLOv5 model for object detection allows us to maintain high accuracy while reducing model complexity and computational demands. Practically, this involves applying regularization pressure to specific network parameters, driving them towards zero throughout the training iterations. This strategy provides us with a more streamlined and efficient network structure [6,11]. Our experimental process is divided into three stages: first, we conduct standard training on the YOLOv5 model to establish a baseline; next, we apply sparsity training to prepare for model pruning; and finally, we perform the pruning process, removing features deemed non-essential, and meticulously adjust the model.

3.2 Lightweight Network

With the rapid advancement of deep learning in computer vision, starting from the advent of AlexNet[10] in 2012, followed by VGG [20] in 2014, and ResNet [5] in 2015, convolutional neural networks have made significant strides in image processing. However, the increasing size and complexity of these networks have escalated hardware requirements, often limiting them to high-performance servers. In order to address the hardware constraints of mobile and embedded devices, lightweight neural networks offer efficient solutions for resource-limited settings, maintaining accuracy while significantly reducing model size and increasing computational speed. In our experiments, substituting YOLOv5's backbone network with these lightweight networks achieved a considerable reduction in model size while preserving accuracy.

MobileNet MobileNetV1 [8], introduced by Andrew G. Howard et al. in 2017, marked a significant shift in convolutional neural network design. Its key innovation is the use of depthwise separable convolutions, which break down standard convolutions into depthwise and pointwise convolutions, greatly reducing parameters and computational load. Unlike traditional convolutions, depthwise convolutions operate separately on each channel of the input feature map, while pointwise convolutions combine these features, akin to the role of fully connected layers. MobileNetV2 [18] further refined the architecture by introducing linear bottlenecks and inverted residual structures, leading to deeper yet more compact and faster networks. MobileNetV3 [7] advanced performance further with hardware-aware Neural Architecture Search (NAS), achieving better performance with less computation.

ShuffleNet [23], proposed by Xiangyu Zhang et al. in 2017, significantly reduces parameters and computational costs through the use of pointwise grouped convolutions. To counter potential information loss from grouped convolutions, it introduces a channel shuffle operation, which rearranges channels after grouped convolutions to exchange information between different groups, enriching feature representation. This unique structural design effectively enhances the network's learning ability, ensuring ShuffleNet maintains high computational efficiency and accuracy while being lightweight, making it particularly suitable for resource-constrained application environments.

GhostNet [3] is an innovative method for building neural networks. It creates feature maps using fewer parameters, reducing computational demands. The key to this method is its Ghost module. This module uses a small number of convolution kernels to produce basic feature maps, and then simple linear operations are applied to create additional *ghost* feature maps. This design concept is crucial for creating lightweight and efficient networks. To integrate the GhostNet architecture into YOLOv5, we modified the original convolution layers, replacing standard convolution layers with Ghost modules. These modules first use fewer filters to generate the main feature maps, then apply simple operations to produce the so-called *ghost* feature maps. This expands the number of feature maps without needing extensive computation.

Table 1. Results on custom THWS dataset(Pruning)

Model Name	Sparcity	Prune ratio	Precision	mAP50[a]	Parameters	Model Size	GFLOPs	Inference Speed(CPU)
Normal Training (Baseline)	–	–	0.8	0.722	7.05M	14.5 MB	15.9	295.7 ms
Sparse Model 1	0.0001	–	0.837	0.741	7.05M	14.5 MB	15.9	302.8 ms
Sparse Model 2	0.0005	–	0.825	0.771	7.05M	14.5 MB	15.9	296.9 ms
Sparse Model 3	0.01	–	0.718	0.658	7.05M	14.5 MB	15.9	295 ms
Prune Model 1	0.0001	30%	0.455	0.304	4.12M	8.111 MB	12.2	246.7 ms
Prune Model 2	0.0001	40%	0.128	0.071	3.36M	6.67 MB	10.8	220.7 ms
Prune Model 3	0.0005	30%	0.835	0.736	4.42M	8.687 MB	13.2	257.8 ms
Prune Model 4	0.0005	50%	0.578	0.44	2.72M	5.429 MB	11.2	219.8 ms
Prune Model 5	0.01	25%	0.717	0.658	5.11M	10.014 MB	13.2	254.1 ms
Fine-tune Model 1	0.0001	30%	0.834	0.788	4.12M	8.302 MB	12.2	240.4 ms
Fine-tune Model 2	0.0001	40%	0.808	0.789	3.36M	6.862 MB	**10.8**	**217.6 ms**
Fine-tune Model 3	0.0005	30%	0.771	0.79	4.42M	8.876 MB	13.2	253 ms
Fine-tune Model 4	0.0005	50%	**0.86**	0.786	**2.72M**	**5.621 MB**	11.2	223.8 ms
Fine-tune Model 5	0.01	25%	0.687	0.679	5.11M	10.208 MB	13.2	258.1 ms

a: mAP50 refers to the Mean Average Precision at an Intersection over Union (IoU) threshold of 50%. It is a common metric used for evaluating the performance of an object detection model at predicting bounding boxes around objects.

Table 2. Results on custom THWS dataset(Replace Backbone)

Model Name	Precision	mAP50	Parameters	Model Size	GFLOPs	Inference Speed(CPU)
Normal Training (Baseline)	0.8	0.722	7.05M	14.5 MB	15.9	295.7 ms
MobileNetv2	0.776	0.661	2.94M	6.4 MB	7.1	243.3 ms
MobileNetv3 Small	0.748	0.663	5.05M	10.6 MB	11.4	227.8 ms
MobileNetv3 Large	0.747	0.674	5.57M	11.6 MB	10.8	265.8 ms
ShuffleNet	0.784	0.66	3.82M	8.1 MB	8.1	212.5 ms
GhostNet	0.759	0.63	3.72M	7.9 MB	8.3	206 ms

4 Experiment

4.1 Dataset

The dataset utilized in the experiment was provided by the Technical University of Applied Sciences Wuerzburg-Schweinfurt (THWS) and consists of various campus images. The reason for using a custom image data set is that we not only compare the performance differences after adjustment of the network model, but also hope that our model can be actually used in some specific THWS campus object detection tasks in the future. This dataset encompasses a variety of natural campus landscapes, buildings, facilities, and various campus events and celebrations. Our team selected 565 images from this dataset, spanning 14 categories for training purposes. These categories include chair, person, tree, robot, door, fire extinguisher, screen, clock, podium, projector, backpack, laptop, bottle, and window. All images were annotated using the LabelImg tool and were divided into training, validation, and test sets at a ratio of 8:1:1. Figure 1 shows the recognition result of some sample images from the dataset. Given the limited size of the dataset, we employed a transfer learning approach using the YOLOv5s as a pre-trained model to expedite the learning process.

Fig. 1. Model predictions on some sample images from the THWS dataset.

4.2 Training, Pruning and Fine-Tuning

Normal Training. We began by employing YOLOv5s.pt as the pre-training model for transfer learning on our THWS image dataset, setting as a baseline. We experimented with various optimizers, including Adam, AdamW, and SGD, each using a uniform input image size of 640 and training for 100 epochs. To reduce data transfer time and reading latency and to enhance GPU utilization, we preloaded and stored image size and label information of the training dataset in memory. This approach resulted in faster model training. For simplicity, we chose AdamW as the optimizer for all subsequent related trainings.

Training with Sparsity. On top of the normal training, we selected different values of the sparse hyper parameter λ from 10^{-5} to 10^{-2} for the channel sparse regularization. To avoid redundant training, all sparse training processes are limited to 50 epochs.

Pruning. After sparse training, we pruned models at different sparsity levels. Our pruning process involved collecting and sorting BN layer weights, setting a threshold to preserve essential features, calculating and displaying the gamma threshold and corresponding pruning ratio, and then adjusting the actual pruning ratio based on the maximum calculated value.

Fine-tuning. We fine-tuned the pruned models, which had a significant reduction in the number of parameters and model size. We made slight alterations to the hyperparameter settings used for fine-tuning compared to normal training. We lowered the initial learning rate from 0.01 to 0.001 and increased the learning rate decay to facilitate faster model convergence. We also made minor adjustments to other hyperparameters related to data augmentation.

4.3 Backbone Replacement

In the experiment involving backbone network replacement, we chose some lightweight neural networks with fewer parameters, simpler network structure, and higher efficiency and reasoning speed to replace the more complex reasoning structure in the original backbone network. These lightweight networks includes MobileNetV2, MobileNetV3-Small, MobileNetV3-Large, ShuffleNet and GhostNet for separate training sessions. Unlike normal training, due to the alteration of the backbone network architecture, we initiated training from scratch instead of using the YOLOv5s model as a pre-training model. Also, we maintained the same hyperparameters as those used in normal training.

5 Discussion

Analysis of the data presented in Table 1 indicates a general decline in mAP50 as model sparsity increases. The pruning experiments reveal that most models

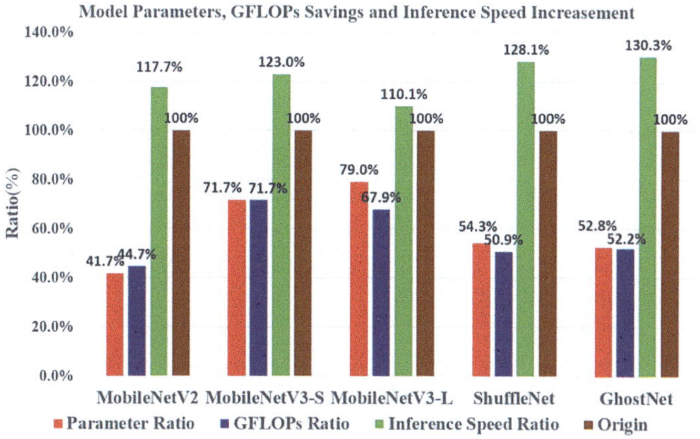

Fig. 2. Comparison of models using lightweight network as backbone with the original model on THWS dataset.

undergo substantial accuracy degradation after more than 30% of parameters are eliminated. However, post-pruning fine-tuning markedly recovers performance. In contrast with the models trained normally, the fine-tuned models exhibit a mere 1.77% drop in average accuracy, yet benefit from a 43% reduction in size, a 20% decrease in GFLOPs, and an 18% average acceleration in CPU inference speed. This underscores fine-tuning as an effective countermeasure to pruning's adverse effects.

The results of replacing the backbone network are similar to those of model pruning. As shown in Table 2 and Fig. 2 for example, after using MobileNetV2 as the backbone network, the model size is reduced by 44.1% compared with the standard model, the inference speed is increased by 17.7%, the GFLOPs are only 44.6% of the baseline, while the mAP50 remains at 0.661. These results demonstrate that lightweight networks can still achieve satisfactory performance in object detection tasks despite the reduced parameters and computational complexity of the model.

In summary, both pruning and backbone network replacement with lightweight networks can effectively reduce the model size, decrease the model complexity as well as improve the inference speed, but may sacrifice the detection accuracy of the model. In view of this, future research should explore how to choose the appropriate pruning ratio and perform related fine-tuning to find a better balance between performance and efficiency. It is also important to investigate how to choose the right backbone network to optimise the trade-off between speed and accuracy in different application scenarios.

6 Conclusions

In this study, we systematically studied the effect of model pruning and replacement backbone network on optimizing the YOLOv5 model. Our experimental results show that by proposing the designed pruning strategy and upgrading the model while significantly reducing the number of parameters, the model can still maintain a high accuracy. Although we achieved certain results, the slight decrease in accuracy also points to the direction of future research. Future research should focus on how to further optimize pruning and fine-tuning strategies without sacrificing detection accuracy to achieve the best balance between performance and efficiency. In addition, it is also crucial to select the most appropriate backbone network for different application scenarios to finely adjust the balance between speed and accuracy.

Ackowledgements. We are grateful to Niko Wörtmann for generously providing us with the necessary datasets for our experiments.

References

1. Bochkovskiy, A., Wang, C.Y., Liao, H.Y.M.: Yolov4: optimal speed and accuracy of object detection. arXiv preprint arXiv:2004.10934 (2020)
2. Guillermo, M., et al.: Detection and classification of public security threats in the Philippines using neural networks. In: 2020 IEEE 2nd Global Conference on Life Sciences and Technologies (LifeTech), pp. 320–324. IEEE (2020)
3. Han, K., Wang, Y., Tian, Q., Guo, J., Xu, C., Xu, C.: Ghostnet: more features from cheap operations. In: Proceedings of the IEEE/CVF Conference on Computer Vision and Pattern Recognition, pp. 1580–1589 (2020)
4. Han, S., Mao, H., Dally, W.J.: Deep compression: Compressing deep neural network with pruning, trained quantization and huffman coding. In: Bengio, Y., LeCun, Y. (eds.) 4th International Conference on Learning Representations, ICLR 2016, San Juan, Puerto Rico, 2–4 May 2016, Conference Track Proceedings (2016). http://arxiv.org/abs/1510.00149
5. He, K., Zhang, X., Ren, S., Sun, J.: Deep residual learning for image recognition. In: 2016 IEEE Conference on Computer Vision and Pattern Recognition, CVPR 2016, Las Vegas, NV, USA, 27–30 June 2016, pp. 770–778. IEEE Computer Society (2016). https://doi.org/10.1109/CVPR.2016.90
6. Hoefler, T., Alistarh, D., Ben-Nun, T., Dryden, N., Peste, A.: Sparsity in deep learning: pruning and growth for efficient inference and training in neural networks. J. Mach. Learn. Res. **22**(1), 10882–11005 (2021)
7. Howard, A., et al.: Searching for mobilenetv3. In: Proceedings of the IEEE/CVF International Conference on Computer Vision, pp. 1314–1324 (2019)
8. Howard, A.G., et al.: Mobilenets: efficient convolutional neural networks for mobile vision applications. arXiv preprint arXiv:1704.04861 (2017)
9. Jithendra, P., Sai, T.V., Mannam, R.K., Manideep, R., Bano, S.: Cognitive model for object detection based on speech-to-text conversion. In: 2020 3rd International Conference on Intelligent Sustainable Systems (ICISS), pp. 843–847. IEEE (2020)
10. Krizhevsky, A., Sutskever, I., Hinton, G.E.: Imagenet classification with deep convolutional neural networks. In: Bartlett, P.L., Pereira, F.C.N., Burges, C.J.C., Bottou, L., Weinberger, K.Q. (eds.) Advances in Neural Information Processing Systems 25: 26th Annual Conference on Neural Information Processing Systems 2012. Proceedings of a meeting held, Lake Tahoe, Nevada, United States, 3–6 December 2012, pp. 1106–1114 (2012). https://proceedings.neurips.cc/paper/2012/hash/c399862d3b9d6b76c8436e924a68c45b-Abstract.html

11. Li, H., Kadav, A., Durdanovic, I., Samet, H., Graf, H.P.: Pruning filters for efficient convnets. In: 5th International Conference on Learning Representations, ICLR 2017, Toulon, France, 24–26 April 2017, Conference Track Proceedings. OpenReview.net (2017). https://openreview.net/forum?id=rJqFGTslg
12. Liu, Z., Li, J., Shen, Z., Huang, G., Yan, S., Zhang, C.: Learning efficient convolutional networks through network slimming. In: Proceedings of the IEEE International Conference on Computer Vision, pp. 2736–2744 (2017)
13. Oquab, M., Bottou, L., Laptev, I., Sivic, J.: Learning and transferring mid-level image representations using convolutional neural networks. In: Proceedings of the IEEE Conference on Computer Vision and Pattern Recognition, pp. 1717–1724 (2014)
14. Redmon, J., Divvala, S., Girshick, R., Farhadi, A.: You only look once: unified, real-time object detection. In: Proceedings of the IEEE Conference on Computer Vision and Pattern Recognition, pp. 779–788 (2016)
15. Redmon, J., Farhadi, A.: Yolo9000: better, faster, stronger. In: Proceedings of the IEEE Conference on Computer Vision and Pattern Recognition, pp. 7263–7271 (2017)
16. Redmon, J., Farhadi, A.: Yolov3: an incremental improvement. arXiv preprint arXiv:1804.02767 (2018)
17. Saha, S., Haque, A., Sidebottom, G.: Transfer learning based efficient traffic prediction with limited training data. In: 2023 IEEE 20th Consumer Communications & Networking Conference (CCNC), pp. 477–480. IEEE (2023)
18. Sandler, M., Howard, A., Zhu, M., Zhmoginov, A., Chen, L.C.: Mobilenetv2: inverted residuals and linear bottlenecks. In: Proceedings of the IEEE Conference on Computer Vision and Pattern Recognition, pp. 4510–4520 (2018)
19. Schwartz, R., Dodge, J., Smith, N.A., Etzioni, O.: Green AI. Commun. ACM **63**(12), 54–63 (2020)
20. Simonyan, K., Zisserman, A.: Very deep convolutional networks for large-scale image recognition. In: Bengio, Y., LeCun, Y. (eds.) 3rd International Conference on Learning Representations, ICLR 2015, San Diego, CA, USA, 7–9 May 2015, Conference Track Proceedings (2015). http://arxiv.org/abs/1409.1556
21. Sindhu, V.S.: Vehicle identification from traffic video surveillance using yolov4. In: 2021 5th International Conference on Intelligent Computing and Control Systems (ICICCS), pp. 1768–1775. IEEE (2021)
22. Yosinski, J., Clune, J., Bengio, Y., Lipson, H.: How transferable are features in deep neural networks? Adv. Neural Inf. Process. Syst. **27** (2014)
23. Zhang, X., Zhou, X., Lin, M., Sun, J.: Shufflenet: an extremely efficient convolutional neural network for mobile devices. In: Proceedings of the IEEE Conference on Computer Vision and Pattern Recognition, pp. 6848–6856 (2018)
24. Zhuang, F., et al.: A comprehensive survey on transfer learning. Proc. IEEE **109**(1), 43–76 (2021). https://doi.org/10.1109/JPROC.2020.3004555

Process Phase Monitoring in Industrial Manufacturing Processes with a Hybrid Unsupervised Learning Strategy

Christian W. Frey(✉)

Fraunhofer Institute of Optronics, System Technologies and Image Exploitation IOSB, 76137 Karlsruhe, Germany
christian.frey@iosb.fraunhofer.de

Abstract. Monitoring production processes, such as those in the chemical industry, involves several key aspects, including identifying unknown process phases, tracking their sequence and duration, and detecting anomalies that may occur within phases. As demonstrated in several industrial applications, the ability of Self-Organizing Maps (SOMs) to detect anomalies, identify and visualize process phases is highly beneficial for comprehensive monitoring and understanding of technical processes. This paper presents a hybrid unsupervised learning strategy (HULS) for monitoring complex industrial processes. Addressing the limitations of SOMs, especially in scenarios with unbalanced datasets and highly correlated process variables, HULS combines existing unsupervised learning techniques to address these challenges. Based on a laboratory batch process, the capabilities of the HULS concept for the detection of unknown process phases are evaluated in comparison to the standard SOM model.

Keywords: unsupervised learning · self-organizing maps · industrial manufacturing process · condition monitoring · anomaly detection

1 Introduction

Industrial batch processes, especially in the pharmaceutical industry, exhibit complex system behaviors, due to the involvement of several chemical reactions in various time-critical process phases. Identifying unknown process phases, tracking their sequence and duration, and detecting anomalies that may occur within phases are key aspects of monitoring such processes. Several concepts in literature use Self-Organizing Maps (SOMs) [11] for process monitoring in various applications. While the approaches in [12] focus on detecting anomalies the concepts in [2,4] expand the utility of SOMs to identify, and monitor process phases. Nevertheless, the application in industrial applications encounters numerous challenges, mainly because SOMs exhibit significant sensitivity to the inherent properties of the training data, such as correlation between the data features [13] and the presence of unbalanced datasets [7]. To overcome these issues, a hybrid unsupervised learning strategy (HULS) has been presented in [5]. In this study, we explore the capabilities of the HULS procedure for identifying process

T. Villmann et al. (Eds.): WSOM+2024, LNNS 1087, pp. 206–215, 2024.
https://doi.org/10.1007/978-3-031-67159-3_23

phases of a laboratory batch process. Section 2 outlines the basic methodologies employed, while Sect. 3 gives an overview of the HULS approach. In Sect. 4, the abilities of the HULS concept for the detection of unknown process phases are evaluated in comparison to the standard SOM model based on a laboratory batch process.

2 Methodologies

2.1 Self-organizing Maps

SOMs are able to map a high-dimensional input dataset $\mathbf{X}$ onto a low-dimensional grid, while preserving the topological structure of the data. From dataset $\mathbf{X} = \{\mathbf{x}_1, ..., \mathbf{x}_L\}$, the input layer receives multidimensional data points with $\mathbf{x}_i \in \mathbb{R}^d \ \forall \ i$, where d is the number of features in $\mathbf{X}$. The output layer is typically a 2D grid (borderless toroid) of size $M \times N$ with $P_{SOM} = MN$ neurons at position $\mathbf{v} \in \mathbb{N} = (1, ..., M) \times (1, ..., N)$. The Best Matching Unit (BMU) is the neuron with index $\mathbf{v}_i^* = \text{argmin}_{\mathbf{v}} \|\mathbf{x}_i - \mathbf{w}_{\mathbf{v}}\|$ whose weight vector $\mathbf{w}_{\mathbf{v}} \in \mathbb{R}^d \ \forall \ \mathbf{v}$ is the most similar to data point $\mathbf{x}_i$, with similarity typically measured by the Euclidean distance. In the k-th step of K training epochs the weights of the BMU and its neighbors are updated by:

$$\mathbf{w}_{\mathbf{v}}(k+1) = \mathbf{w}_{\mathbf{v}}(k) + \alpha(k) h_{\mathbf{v}^*\mathbf{v}}(k)(\mathbf{x}_i - \mathbf{w}_{\mathbf{v}}), \tag{1}$$

where $\alpha(k)$ is the learning rate and $h_{\mathbf{v}^*\mathbf{v}}(k)$ is the neighborhood function. Both functions can be applied in several ways [11], however they should be decreasing with k. With α_0 as initial learning and σ_0 the initial neighborhood radius in the following experiments $\alpha(k)$ and $h_{\mathbf{v}^*\mathbf{v}}(k)$ are defined as:

$$\alpha(k) = \alpha_0 e^{-\frac{k}{K}} \quad (2) \quad , \text{and } h_{\mathbf{v}^*\mathbf{v}}(k) = e^{-\frac{\|\mathbf{v}^* - \mathbf{v}\|}{\sigma(k)}} \text{ with } \sigma(k) = \sigma_0 e^{-\frac{k}{K}} . \quad (3)$$

Mapping quality of a trained SOM cannot be measured by a single energy function therefore several metrics are available in the literature [3]. In this study, quantization error E_Q and topographic error E_T are chosen:

$$E_Q = \frac{1}{L} \sum_{i=1}^{L} \|\mathbf{x}_i - \mathbf{w}_{\mathbf{v}^*}\| \quad (4) \qquad , \text{and} \quad E_T = \frac{1}{L} \sum_{i=1}^{L} \delta_i. \quad (5)$$

E_Q measures the mean error between the data points $\mathbf{x}_i$ and $\mathbf{w}_{\mathbf{v}^*}$, while E_T measures how well spatial relationships in $\mathbf{X}$ are preserved in the lattice after training. In a perfect SOM, the BMU and the second BMU should be adjacent ($\delta_i = 0$). If they are not, it is considered as a topographic error ($\delta_i = 1$).

2.2 Unified Distance Matrix

The Unified Distance Matrix (UM) is a valuable tool for visualizing and interpreting SOMs, providing insight into the underlying structure and clusters within the data [14]. With $\mathbf{v} = (m, n)$ as the position of the individual neuron and $\mathbf{p} = (m-1, ...m+1) \times (n-1, ..., n+1)$ as the position of the neighbors in the lattice, the distance $U_\mathbf{v}$ is calculated by:

$$U_\mathbf{v} = \frac{1}{8} \sum_{p \neq v} \|\mathbf{w_v} - \mathbf{w_p}\| . \tag{6}$$

The UM visualization provides evidence for the effectiveness of SOMs, as neurons with highly similar weight vectors are typically situated near each other, forming valleys, and ridges where they differ. In the context of monitoring production processes the valleys can be interpreted as distinct process phases while ridges mark transitions between the phases [4].

2.3 Instantaneous Topological Map

SOMs suffer from severe difficulties when the training dataset $\mathbf{X}$ is unbalanced with respect to the expected number of clusters [7] or the features are highly correlated [13]. There are several approaches to overcome these problems [13], one of which is the Instantaneous Topological Map (ITM) [10]. The input layer receives the data points $\mathbf{x}_i \in \mathbb{R}^d \ \forall \ i$ from the dataset $\mathbf{X}$, while the output layer holds P_{ITM} neurons with an associated weight vector $\mathbf{s}_r \in \mathbb{R}^d \ \forall \ r$ and a set of edges Q_j connecting the neighboring neurons. An ITM has no fixed number of neurons P_{ITM}. Creation of neurons is based on the relation between the BMU j^* and the second BMU j^{**}, which can be determined, for example, by the Euclidean distance:

$$j_i^* = \text{argmin}_\mathbf{j} \|\mathbf{x}_i - \mathbf{s}_j\| , \ j_i^{**} = \text{argmin}_{\mathbf{j} \neq \mathbf{j}^*} \|\mathbf{x}_i - \mathbf{s}_j\| . \tag{7}$$

The ITM method does not project the dataset onto a predefined lattice with a fixed number of neurons. Instead, ITM maps the data to a flexible number P_{ITM} of neurons with the topological relationships represented by Q_j edges between neurons. The length of these connections indicates the topological proximity of the neurons involved. Mapping sensitivity is regulated by the threshold parameter β, which corresponds to the radius of a sphere surrounding a neuron. A neuron is created whether or not x_i falls within this sphere. Starting with P_{ITM}=2 randomly initialized neurons, a new neuron $\mathbf{s}_{P+1} = \mathbf{x}_i$ with an edge between $\mathbf{s}_{j^*}$ and $\mathbf{s}_{P+1}$ is created, if the following condition is met [10]:

$$\|\mathbf{x}_i - \mathbf{s}_{j^*}\| > \beta \wedge \|\mathbf{s}_{j^*} - \mathbf{s}_{j^{**}}\| < \|\mathbf{x}_i - \mathbf{s}_{j^*}\| . \tag{8}$$

Immediately after a node is created j^* and j^{**} are determined again based on $\mathbf{x}_i$. If the following condition is true, an unnecessary j^{**} and the corresponding edges are removed:

$$\|\mathbf{s}_{j^*} - \mathbf{s}_{j^{**}}\| < 0.5\beta \tag{9}$$

The ITM procedure is incremental - the map is updated instantly by sequentially processing data points x_i. Map training is only controlled by β, where a smaller β leads to a higher number P_{ITM} in the map. This aspect can also be interpreted in the context of resampling [7], where the weight vectors of the neurons $\mathbf{s}$ reflect the resampled dataset $\mathbf{X}$.

3 Hybrid Unsupervised Learning Strategy

To address the previously mentioned limitations of SOMs in processing unbalanced or highly correlated training data, we propose a hybrid approach that combines the capabilities of SOMs and ITM. While ITM is also a topology preserving neural network it can handle correlated training data efficiently and resample the training data in an intuitive manner [1]. The output of an ITM is a set of neurons with associated weight vectors $\mathbf{s}_j$ and associated edges Q_j. However, unlike SOMs, ITMs do not map the high-dimensional input dataset $\mathbf{X}$ onto a 2D grid and therefore they lack an effective and intuitive clustering mechanism similar to, for example, UM based techniques. This is a key aspect considering the challenge of identifying and monitoring unknown process phases. Therefore the proposed HULS concept is based on the following idea: In the first step, the ITM algorithm is applied to the training dataset $\mathbf{X}$. Based on the threshold parameter β of the ITM procedure, the generation of neurons is controlled. The resulting set of ITM neuron weights $\mathbf{S}$ can be understood as the resampled dataset of $\mathbf{X}$. In the next step, the set $\mathbf{S}$ is then used as the training dataset for the SOM procedure. Based on the trained SOM model, the subsequent steps of UM and WT can be performed for automated cluster identification (Algorithm 1).

3.1 Mechanism for SOM Map Size Adaptation

Predefining the map size of a SOM imposes a significant constraint on the ability to provide a correct topological representation of the data, and may lead to incorrect interpretation [14]. Typically, a set of SOMs with predefined sizes is generated and evaluated based on subjective criteria, which is a major limitation of conventional SOMs and their industrial applications. To overcome this limitation, numerous approaches have been presented in the literature [8], which mainly differ in the mechanism of how neurons are added to the grid and whether they maintain a 2D grid structure. Inspired by the growing grid (GG) model [6], we developed a simplified mechanism for map size adaptation. In contrast to GGs, where neuron insertion is triggered by the frequency on how often a neuron has become a BMU, the proposed mechanism relies on the neighborhood function $h_{\mathbf{v}^*\mathbf{v}}(k)$ and the BMU position $\mathbf{v}_i^*$ in the grid. Figure 1 illustrates the mechanism. Considering the k-th training step, which processes

Algorithm 1. Hybrid Unsupervised Learning Strategy $HULS(\mathbf{X})$

Input: $\mathbf{X}$, β, K, α_0, σ_0, ϕ

1: Instantaneous Topological Mapping $ITM(\mathbf{X})$
2: Self Organizing Map $SOM(\mathbf{S})$
3: Unified Distance Matrix $UM(SOM)$
4: Watershed Transformation $WT(UM)$

Output: ITM, SOM, UM, WT, $\mathbf{C}$, $\mathbf{v}_C$

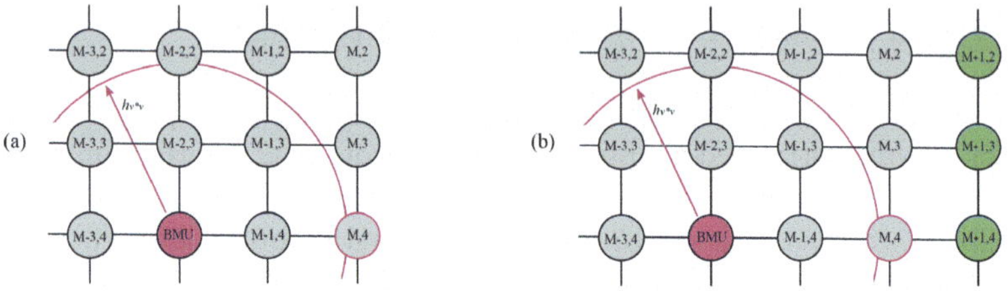

Fig. 1. SOM map adaptation: With $h_{\mathbf{v}^*\mathbf{v}}(k)$ the BMU neighbors are determined. If a neighbor is a border neuron (a), a column/row of border neurons is inserted (b).

the i-th data point $\mathbf{x}_i$, like in standard SOMs the BMU and the neighborhood radius $h_{\mathbf{v}^*\mathbf{v}}(k)$ are calculated. Before adapting the weights of the neighboring neurons with Eq. 1, their position in the grid is checked. Neuron insertion is triggered, if the neuron is within the radius $h_{\mathbf{v}^*\mathbf{v}}$ and if the neuron is at the border of the grid (left, right, top, bottom). In this case a complete row or column of new neurons is added to the affected grid border. As the neighborhood function $h_{\mathbf{v}^*\mathbf{v}}(k)$ decreases over the training iterations, the growth of the grid is limited. The proposed approach keeps a rectangular grid with "unused" neurons available at the borders, which can be further processed by the UM transform.

3.2 Automated Segmentation of the U-Matrix

As introduced in Sect. 2.2, the UM enables the visualization of the topological relationships in a trained SOM. In the context of industrial production processes, the valleys can be interpreted as distinct phases of the process, while the ridges mark transitions between them [4]. Although it is possible to manually segment a UM, this approach becomes impractical as the grid size or the structural complexity increases. To automatically segment the UM, clustering algorithms from the image processing domain, such as the Watershed Transformation (WT), can be utilized [15]. To improve the quality of the WT for UM clustering, a three-step procedure is proposed: First, to remove noise and artifacts, the UM is filtered by a 2-D Gaussian kernel with standard deviation σ_{kernel}. To avoid over-segmentation and to preserve obvious segments an H-minima transform [9] is applied, suppressing values in the UM below a given threshold ϕ. The output is the matrix $\mathbf{H}$ with the elements scaled e.g. to [0, 254]. Finally, the non-recursive WT approach is applied [15]. In summary, the proposed approach with given σ and ϕ maps each neuron $\mathbf{v}$ to a cluster c, where C is the set of identified clusters and $c \in \mathbb{N} = (0, 1, ..., |C|)$. Experiments showed, that the segmentation is mainly controlled by the margin ϕ, while it is less sensitive to the kernel width σ_{kernel}. Choosing inappropriate values of ϕ will lead to over-/under-segmentation, i.e. an inappropriate number of process phases identified. To set the margin ϕ automatically, the following considerations are made: Starting with $\phi = 1$, the number of clusters will be $|C| = max$. Increasing ϕ will reduce $|C|$ until $|C| = 1$. If the topological mapping of the SOM was successful, i.e. there are clusters in the UM, there should be a stable local minimum where $|C|(\phi)$ is constant over a certain interval ϕ of length d:

$$\arg\max_{\phi}(\arg\max_{d}\{i \in 0, \ldots, d \mid \; |C|(\phi) = |C|(\phi + i)\}) \tag{10}$$

with ϕ in the range of 1 to 254. This assumption may not be universally valid, but it leads to a simple brute force approach for determining the margin ϕ.

4 Experiments

Industrial batch processes, particularly in the pharmaceutical industry, exhibit complex system behaviors, due to the involvement of multiple chemical reactions in various time-critical process phases. Considering these premises, the capabilities of the proposed HULS concept will be evaluated based on an exemplary laboratory batch process.

4.1 Laboratory Batch Process and Data Acquisition

Figure 2 shows the simplified flowchart of the laboratory batch process. The process includes two liquid tanks, with tank B02 functioning as a supply reservoir for the process liquid. The outlet of tank B02 is equipped with a pressure sensor (P) to estimate the fill level, a flow meter (F), and an electro-pneumatic valve (Z) to regulate the flow rate of the liquid feed into the reaction tank B01. The level in B01 is measured by an ultrasonic level meter (L). Once B02 is filled to a defined level and after a product-specific reaction time, the liquid is transferred to a succeeding process unit by means of pump P01 with speed control (D) to adjust the flow rate. It should be noted that for the experiments, to illustrate a batch process behavior, the pre-product is returned to the tank B02. For the following experiments, a dataset for training and validation was recorded. Figure 3 (a) shows the time course of the five process values (P,F,L,D,Z) over six batches, with batches T1 through T4 used for training, and batches N1 and N2 used for validation. Additionally shown is the time course of the target phases c_{Target}. Note that the c_{Target} is used for comparison only and is not used for training. The process values are normalized and time reference is anonymized due to confidentiality restrictions.

4.2 Discovering Unknown Process Phases

As introduced in Sect. 2, the combination of SOMs with UM and WT provides a compelling method for clustering a dataset in an unsupervised manner. To evaluate the ability of the proposed HULS concept to identify unknown process phases, a comparative evaluation against the standard SOM procedure is conducted, based on the dataset presented in Sect. 4.1. As far as possible, a harmonized training setup was chosen with the training parameters are set to $K = 500$ iterations, $\alpha_0 = 0.02$ as initial learning rate, $\sigma_0 = 2$ as initial influence

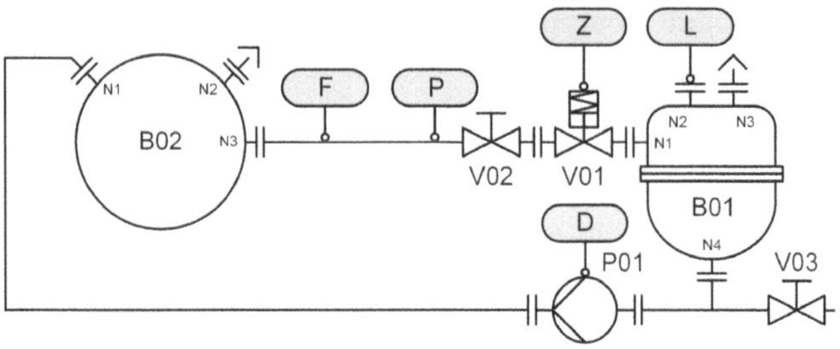

Fig. 2. Simplified flow chart of the laboratory batch process with measured values flow (F), pressure (P), level (L), and control values pump speed (D), and valve position (Z)

radius, and $\sigma_{kernel} = 1.6$ as UM filter parameter. The additional HULS parameter β is set to $\beta = 0.01$. The map size for the SOM model is set to a fixed size of $P_{SOM} = 55 \times 55$, while for the HULS procedure it is set to an initial size of $P_{HULSStart} = 11 \times 11$.

Table 1 provides, as a first result, a comparison of the key quality measures for the two algorithms after training. Comparing the quantization error E_Q gives an idea of how well the two models perform in covering the training dataset. In our experiments, the proposed HULS method performs significantly better than the standard SOM. A similar result is obtained when looking at the topographic error E_T - again, the HULS procedure outperforms the standard SOM. A high topological error E_T indicates that the topological mapping is insufficient. This effect can be verified by examining the UM and the WT of both models. As shown in Fig. 4(a), the SOM was only able to identify $|C_{SOM}| = 4$ distinct process phases instead of the expected the $|C_{Target}| = 5$. A closer look at the UM reveals that the SOM is not able to generate an adequate topological representation and accordingly the process phases are not reliably recognized by the WT. In contrast, the HULS concept (see Fig. 4(b)) successfully and reliably identifies $|C_{HULS}| = 5$ process phases. The SOM model is not suited for on-line monitoring of the process phases. This is particularly evident when looking at the time courses of the identified process phases in Fig. 3(b/c). While the HULS concept correctly assigns process phases $c = 2$ and $c = 3$, the SOM concept skips the short phase $c = 3$ and assigns it to a prolonged phase $c = 2$. This is due to the fact that SOMs cannot handle unbalanced datasets very well.

Another interesting aspect of the HULS concept is the automatic adjustment of the map size. Before training of the map the HULS map size was set to

Table 1. Comparison of mapping quality for standard SOM and HULS procedure

	Quantization Error E_Q	Topographic Error E_T
SOM	0.144	11.39 %
HULS	0.061	5.72 %

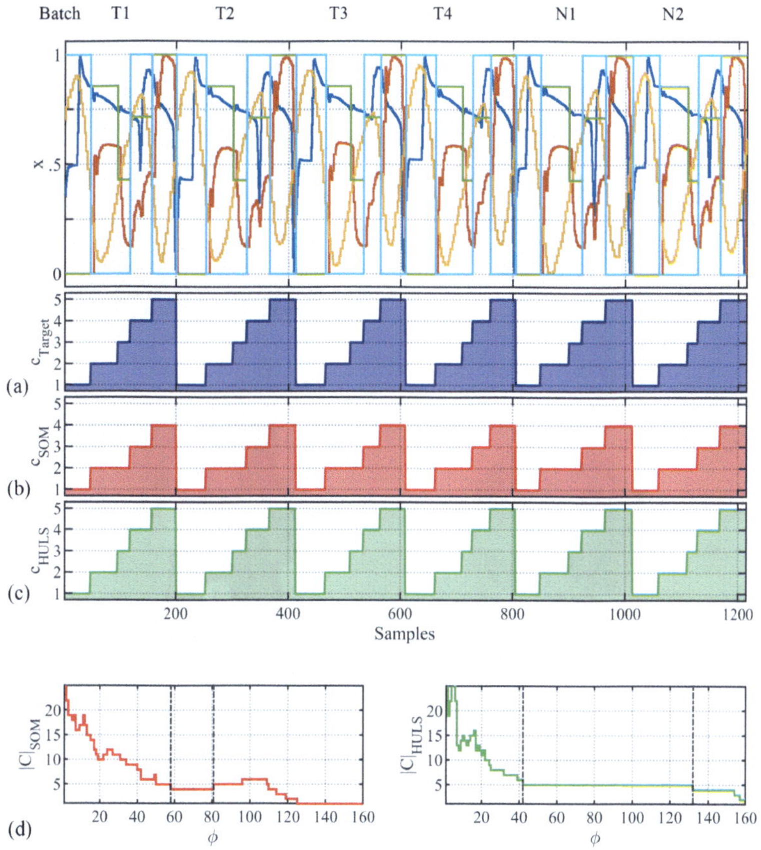

Fig. 3. (a) Recorded data $\mathbf{X}$ with batches for training T1-4, evaluation N1-2, and target process phases c_{Target}. (b) Identified phases by the SOM c_{SOM} and (c) HULS procedure c_{HULS}. (d) Course of $|C|$ versus H-minima margin ϕ for SOM and HULS.

$P_{HULSStart} = 11 \times 11$. During the training phase, the grid evolved to a rectangular grid size of $P_{HULSEnd} = 73 \times 71$, which is a total number of 5183 neurons, but only 2513 neurons are used. Compared to the standard SOM with a fixed number of 2601 used neurons, this very good result proves that the developed map adaptation mechanism is very robust and efficient. As can be seen in Fig. 4, the main effect of map adaptation is to allow the map to grow when needed (b), overcoming the constriction effect of borderless SOMs (a).

In the experiments performed, both the HULS and the SOM model were automatically segmented using the procedure presented in Sect. 3.2. Figure 3(d) visualizes the number of identified clusters as a function of the H-minimum margin ϕ. Starting with phi=1 gives the maximum number of clusters. By successively increasing ϕ, the number of clusters $|C_{SOM}|$ is successively reduced. Comparing the progression $|C_{SOM}|$ between the two methods, it can be seen that the HULS

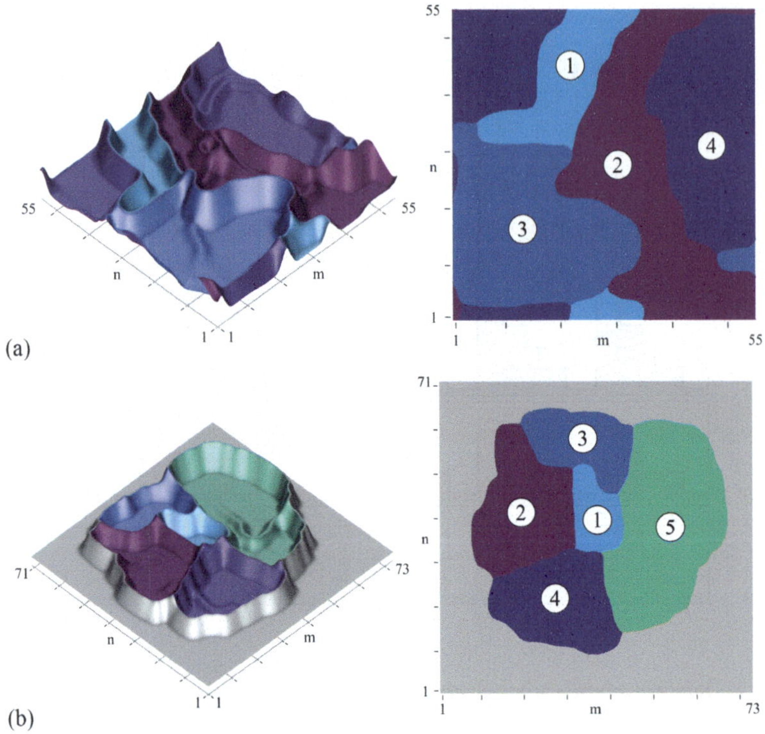

Fig. 4. 3D/2D Visualization of UM and WT (a) for standard SOM with $|C_{SOM}| = 4$ and (b) for proposed HULS procedure with $|C_{SOM}| = 5$ identified process phases.

model gives a stable plateau $|C_{HULS}| = 5$ in a range of $\phi = [42, 132]$. In contrast, the SOM model shows only a short plateau with $|C_{SOM}| = 4$ in the range of phi $= \phi = [59, 81]$. Based on this result, the H-minimum margin was set to $\phi_{SOM} = 70$ and $\phi_{HULS} = 87$, respectively.

5 Conclusions

Achieving cost-effective operation of production systems requires continuous monitoring of plant functionality. The ability of SOMs to detect anomalies, identify and visualize process phases is highly beneficial for comprehensive monitoring and understanding of technical processes. However, SOMs have certain limitations, especially when dealing with unbalanced datasets and highly correlated process variables. When comparing the experimental results in Sect. 4.2, the efficiency of the developed HULS concept becomes apparent. Even under challenging conditions, the HULS method is able to reliably identify the unknown process phases. With the implemented mechanisms for SOM map size adaptation and automated process phase identification, the HULS concept can be applied in industrial applications. Currently, the concept is being tested and evaluated in various industrial domains. The applications range from process industry, power plants to monitoring of aggregates in the chemical and pharmaceutical industry.

References

1. Barreto, G.D.A., Araujo, A.F.R., Ritter, H.J.: Self-organizing feature maps for modeling and control of robotic manipulators. J. Intell. Rob. Syst. **36**, 407–450 (2003)
2. von Birgelen, A., Niggemann, O.: Using self-organizing maps to learn hybrid timed automata in absence of discrete events. In: 2017 22nd IEEE International Conference on Emerging Technologies and Factory Automation (ETFA), pp. 1–8 (2017)
3. Erwin, E., Obermayer, K., Schulten, K.: Self-organizing maps: ordering, convergence properties and energy functions. Biol. Cybern. **67**, 47–55 (1992)
4. Frey, C.W.: Monitoring of complex industrial processes based on self-organizing maps and watershed transformations. In: 2008 IEEE International Conference on Industrial Technology, pp. 1041–1046 (2008)
5. Frey, C.W.: A hybrid unsupervised learning strategy for monitoring complex industrial manufacturing processes. In: IECON 2023 - 49th Annual Conference of the IEEE Industrial Electronics Society, pp. 1–8 (2023)
6. Fritzke, B.: Growing grid - a self-organizing network with constant neighborhood range and adaptation strength. Neural Process. Lett. **2**, 9–13 (1995)
7. He, H., Garcia, E.A.: Learning from imbalanced data. IEEE Trans. Knowl. Data Eng. **21**, 1263–1284 (2009)
8. Hung, C., Wermter, S., Chi, Y.L., Tsai, C.F.: An adaptive growing grid model for a non-stationary environment. Neurocomputing **550**, 126405 (2023)
9. Ismail, N., Zaini, T.R.M., Jaafar, M.Z., Pin, N.C.: H-minima transform for segmentation of structured surface (2016). https://api.semanticscholar.org/CorpusID:63857258
10. Jockusch, J., Ritter, H.J.: An instantaneous topological mapping model for correlated stimuli. In: IJCNN 1999. International Joint Conference on Neural Networks. Proceedings (Cat. No.99CH36339), vol. 1, pp. 529–534 (1999)
11. Kohonen, T.: Essentials of the self-organizing map. Neural Networks Official J. Int. Neural Network Soc. **37**, 52–65 (2013)
12. Li, N., Jiang, K., Ma, Z., Wei, X., Hong, X., Gong, Y.: Anomaly detection via self-organizing map. In: 2021 IEEE International Conference on Image Processing (ICIP), pp. 974–978 (2021)
13. Ruppert, D.: The elements of statistical learning: data mining, inference, and prediction. J. Am. Stat. Assoc. **99**, 567–567 (2004)
14. Ultsch, A.: Maps for the visualization of high-dimensional data spaces (2003). https://api.semanticscholar.org/CorpusID:6808278
15. Vincent, L.M., Soille, P.: Watersheds in digital spaces: an efficient algorithm based on immersion simulations. IEEE Trans. Pattern Anal. Mach. Intell. **13**, 583–598 (1991)

Knowledge Management in SMEs: Applying Link Prediction for Assisted Decision Making

Steven Lehmann[1(✉)], Jörg Schließer[1], Sandra Schumann[1], Heiner Winkler[2], and Iren Jabs[2]

[1] Institute for Applied Informatics at Leipzig University, 04109 Leipzig, Germany
steven.lehmann@infai.org

[2] Institute for Management and Factory Systems, Chemnitz University of Technology, 09125 Chemnitz, Germany

Abstract. Knowledge Graph Completion via Link Prediction (LP) is an important topic for current studies. In this paper, we present a state-of-the-art method to support decision-making within the context of project planning in small and medium-sized enterprises. A particular focus is on creating a suitable ontology based on real company data, which serves as a basis for LP. This work introduces a novel approach to assist managers in decision making for better use of human resources. By accurately predicting which teams are suitable to work on a new project, we aim to significantly reduce the time needed for manual assignment.

1 Introduction

Digitalization is an important aspect of the transformation towards Industry 4.0, which was initiated by the German government in 2011 with the goal to provide companies with guidelines to stay competitive in a quickly evolving environment [15]. This concept is extended to Industry 5.0 which focuses on the human-centered introduction of Artificial Intelligence (AI) and puts people at the center of manufacturing [12]. Digitalization, AI and human centeredness are challenging for small and medium-sized enterprises (SMEs) in particular, as they often only have limited resources and prior knowledge [11].

The topic of this work is to design a method using AI to assign teams to incoming projects. This is done for a single company, with a unique data structure and workflow. Therefore, while the preprocessing steps of the data are specific to this project, the overall approach and resulting methodology can be transferred to similar tasks. It supports managers in decision making for optimized use of human resources. On one hand, it relieves the burden on managers, on the other, employees in the teams are assigned to projects that best suit their interests and competences.

Specifically for the use case in this company, there are almost 300 previous industrial- and scientific projects with a huge quantity of analogous documentation. The company is structured in departments, which are spanned across two

T. Villmann et al. (Eds.): WSOM+2024, LNNS 1087, pp. 216–225, 2024.
https://doi.org/10.1007/978-3-031-67159-3_24

different sites. A unified language between departments is missing, communication can vary depending on the person in charge of the department.

In order to solve the task of automating the project assignment task, we identified Knowledge Graphs (KG) as a suitable method to compose and visualize the current knowledge. KGs store knowledge in the form of triples, containing head, relation and tail (h, r, t) each. Using a KG and its inherent structure, we model the present assignment problem as a Link Prediction (LP) task. This paper is structured as follows. In Sect. 2 the related work is summarized. Section 3 describes the approach and methodology of this work. Section 4 presents the results, before a conclusion is drawn and an outlook for future work is given in Sect. 5.

2 Related Work

Link Prediction describes the task for predicting edges in a network, that could occur in the future [10]. In the domain of KG, knowledge graph embeddings models (KGEMs) are typically used to perform the task of LP to either predict the head element for a given relation-tail pair or the tail element for a given head-relation pair. The transformation method assigns each element of the KG with a randomized representation in the vector space, called knowledge graph embeddings (KGE) [16]. The KGEM introduces an evaluation function to assess the plausibility of each triple. Using optimization algorithms, the vectors are then adjusted to enhance the overall plausibility [7]. A KGEM typically consists of 4 integral parts: the interaction model, the training approach regarding negative samples, the loss function and the usage of inverse relations [2].

The interaction model as central part of a KGEM is responsible for computing this plausibility measure. Over the years, different classifications of interaction models were proposed. Following the categorization of Wang et al. [21], the interaction models can be classified into translational distance based and semantic matching based models. While translational based models like TransE [6] or RotatE [17] utilize distance functions to measure the plausibility between two entities, semantic matching models like MuRE [4] use the semantic attributes of the elements of the knowledge graph to assess the plausibility via similarity-based scoring [21]. Based on the implemented method, these interaction models have different advantages and disadvantages with regards to modelling types of relations like symmetry and composition, as well as computation cost.

There are two main training approaches for creating negative samples, the local closed world assumption (LCWA) [9] and the stochastic local closed world assumption (sLCWA) [23]. Because triples in KGs are assumed to be correct, negative samples are created to explicitly give weight to the fact that triples can be false. While sLCWA samples a subset of negative triples based on different sampling algorithms, with LCWA a negative triple is created for every any triple in the dataset.

Loss functions are a part of every Neural Network (NN). In the context of KGEMs, loss functions directly impact the performance. There are several different loss functions, all which are suited for different use cases regarding LP.

The explicit use of inverse relations is the last integral part of the LP pipeline. By using inverse relations, a set is created by inverting all triples present in the dataset, resulting in the doubling of the embedding space.

Typically, LP based on KGEMs is evaluated using rank-based metrics such as *hits@k*, where the term describes the percentage of test triples that have been ranked in the top k triples. Different configurations of the different parts yield various results depending on the dataset [2]. A variety of different KGEMs have been proposed over the past few years. Ali et al. [2] composed an extensive experiment, using 21 KGEMs on 4 benchmark datasets, to systematically assess which models and methods are best suited for LP. While not explicitly stated, this work is constrained on the transductive use-case. A transductive setup with regard to LP means that all entities of the KG are known at the time of training [1]. They highlighted the importance of fine-tuning the KGEMs, as this can lead to significant improvements, even for poorly suited models.

There is a significant lack of published work on the topic of LP in SMEs. Li et al. [14] applied an LP task in the context of supply chains in SMEs. Apart from this application domain, there are use cases that implement recommendation systems based on LP [22]. Yang et al. [24] studied evaluation methodologies for knowledge graph completion under the open-world assumption. They stated, that only focusing on typical evaluation metrics such as *hits@k* has shortcomings with regards to properly reflecting model improvements and identifying the best model when comparing models on the complete KG versus on a subset.

Recent developments focus on improving semi-inductive LP with transformers [20] and pre-trained language models (PLMs) [13].

3 Methodology

The given data consists of several Excel sheets describing the departments, teams with their competences, and previous and ongoing projects. The company consists of 9 departments and 23 teams. 299 projects are contained in the digitized data, with 1–6 teams assigned each. Three team have never been assigned a project and another three worked only on a single project. Whether this is due to missing data, or because the teams have been established recently is unknown. An overview over the number of assigned projects for each team is visualized in Fig. 1. In total 375 teams were assigned to the 299 projects. The team with the most assignments worked on 47 projects. This distribution is mostly relevant for the interpretation of the results discussed in Sect. 4.

While data for the departments and teams is complete, the project data is sparse when it comes to descriptors such as title, keywords, and textual descriptions. Needed competences for each project are completely missing, which is unfortunate, as this would give a direct link between projects and teams. For this work, we therefore only use the other descriptors featured in the data to train a NN for LP. The company has planned to retroactively add competences to projects where possible, which will then be used for future work. The only required property for a project to be included in the dataset is the list of required

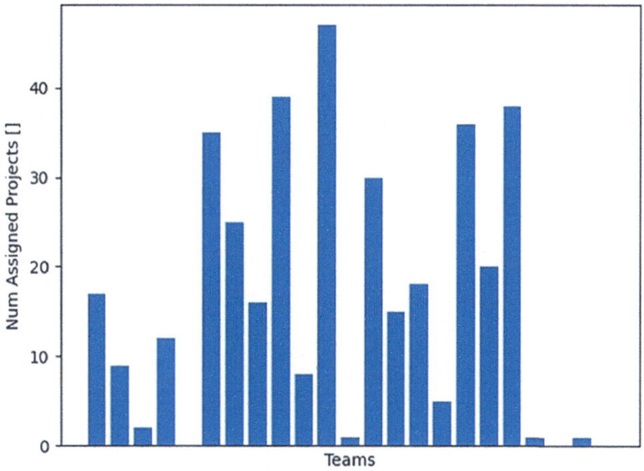

Fig. 1. Bar chart of the amount of project assignments for the 23 teams present in the dataset

teams, which is present for all projects. Removing projects with other missing fields would cause a too large reduction in the dataset.

For creating the KG we used the process proposed by Tamašauskaitė et al. [18]. Using a bottom-down approach, first an ontology based on the entities and relations described was defined. After that, the extraction of entities, relations and attributes corresponding to the concepts in the ontology took place. After the extraction process, the data was stored in an Resource Description Framework file (RDF). The RDF consists of 7418 triples, based on 10 relations and their 10 corresponding inverse relations, as well as a subset of project descriptions which are stored as properties. The ontology is visualized in Fig. 2. We implemented an 80%/10%/10% split for the training, validation and testing data, but only applied the split on the relation to predict and its inverse. All triples with other relations were added to the training set. One advantage of this approach is that triples with relations that are not used for LP would be wasted in the test set. Another is that the method to split new projects will be identical in the application. This resulted in a split of roughly 98%/1%/1%. For this use-case two aspects have to be considered. (1) A relation has to be in the same dataset as its inverse and (2) the evaluation metrics for predicting a tail will differ for the relevant relation called "assignedTo", which is part of the triple "Project assignedTo Team" and the inverse called "worksOn". This is because the inverse relation has, for most heads in the testing dataset, several true tails in the training set. Those true tails are then preferred guesses in the validation and testing steps.

The implementation of the KGEM is done with the PyKEEN [3] framework. The model training is done locally on a GeForce RTX 4050 GPU. In this work, transductive LP is used, meaning that the model has to be retrained whenever a new entity is added. This approach is applicable due to the small data size.

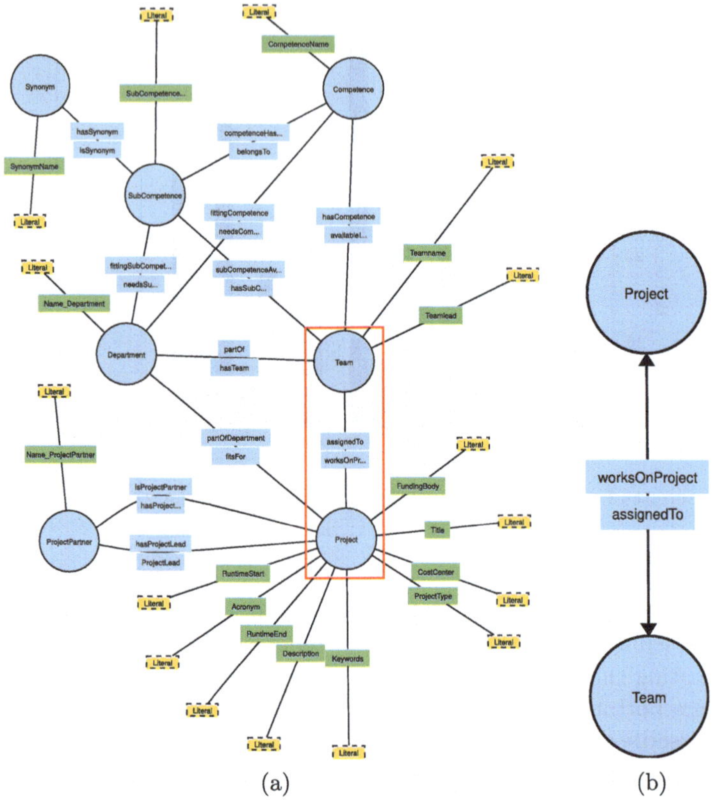

Fig. 2. Ontology of the available data

When a new project is added, the resulting workflow is the following: The provided description is added to the training data, the model then gets retrained before guessing the relation "assignedTo" for the new project entity. The resulting scores for each team can then be classified depending on whether the score indicates a good assignment fit. An ablation study, which tests a variety of combinations of models, loss functions, and other hyper-parameters, is performed. A brief overview over insights gained from the study is presented in Sect. 4, along with detailed metrics of the best performing models.

4 Results

In this section we provide the results for our LP task. As the hyper-parameters have been determined experimentally in the ablation study, we will not discuss the KGEM settings in detail. The results for trained models are shown in Table 1. For this purpose, we created 3 different data splits and trained each model 3 times on each split. Early-stopping is implemented, because tests without did not show significant improvements and, while only a few minutes long, the training duration is supposed to be minimal for the application in real-time. To improve comparability between the models, random seeds are specified, so that e.g., in every trial for TransE the same seed is used. Each row lists the mean for each model. We decided to focus on *hits@k* as our evaluation metrics, with *hits@*1 as the metric to optimize during training. During the optimization process there was little difference between the results for various values for k.

Several models are suited for this LP task, adding more importance to the training time. TransE with MarginRanking Loss is by far the fastest model and achieves decent results but still performs significantly worse than other models. TransE with Softplus Loss loses most of its speed advantage but is more competitive in terms of predictive accuracy.

RotatE achieves the best results prediction wise, but needs a comparatively very long training duration. MuRE achieves results that are competitive with all other interaction models except for RotatE and demonstrates exceptional speed when using BCEWithLogits as loss function. Other notable models include ConvE [8], TuckER [5] and ComplEx [19]. The results from the ablation study also showed, that the configuration of the LP pipeline significantly impacts the prediction results depending on the chosen interaction model or loss function. There are differences in the distribution of the best guesses for the test data among the interaction models. RotatE suggests teams 9 and 11, which are the most present in the dataset (see Fig. 1) with the highest frequency. For both MuRE models, team 7 is overrepresented. The probability of randomly selecting one of those teams based on the distribution in the full dataset is lower than the probability of selecting them during the testing stage. Table 2 presents a more exemplary insight into two different partial results for a given model. In this case MuRE with BCEWithLogits Loss under the sLCWA was used. The example illustrates a perfect score, as for "Project A", where the model predicted the teams that were assigned to the project received the highest scores with a

Table 1. *Hits@k* metric for various models. Each value represents the mean over several trial runs with variations in data split. Training time for the model is given in seconds.

Model	Loss Function	Duration[s]	Hits@1	Hits@3	Hits@5	Hits@10
TransE	MarginRanking	13	25.94	47.53	61.65	80.13
TransE	Softplus	99	42.49	76.36	87.70	96.07
RotatE	BCEWithLogits	2195	61.39	85.45	93.87	96.71
MuRE	Softplus	466	51.15	81.26	90.02	97.99
MuRE	BCEWithLogits	22	47.39	79.71	91.75	97.18
ConvE	CrossEntropy	126	46.53	72.93	85.42	96.53
TuckER	MarginRanking	1241	53.02	82.29	93.81	97.78
ComplEx	BCEAfterSigmoid	204	39.66	71.40	85.11	92.45

significant gap to the next best team and an even larger gap after. For "Project B" the correct team is only ranked 5 based on the scores, but the distance to other teams is relatively low. A positive aspect of those results is that a variety of teams receive the highest scores. The highest score achieved by a single predicted relation in the test set during this run was ~ 5, which leaves some uncertainty for smaller scores. For most projects, if a team was assigned a score > 2, the best fitting team was correctly predicted. In the few exceptions, besides erroneous model behavior, it is possible that a good fitting team wasn't assigned to a project because of missing capacities or other reasons not reflected in the data.

Table 2. Detailed presentation of results from a given model trained with MuRE as interaction model. The scores with predicted relations that are contained in the test data are in bold with green background.

	Project A		Project B	
Rank	Team ID	Score	Team ID	Score
1	3	**2.65**	6	1.62
2	5	**2.20**	10	1.56
3	0	1.00	8	1.22
4	14	−0.51	11	1.01
5	4	−0.62	7	**0.93**

To get an insight into behavior of the models when a new project is to be assigned, we created several dummy projects with various settings. The dummy projects are supposed to include (1) a project that is designed for a single specific team, (2) a project that is designed for a specific team but only via the competences, (3) a project that is designed as a mediocre fit to some teams, and (4) a project with completely unrelated data. The idea behind (2) is to test whether the model learns to transfer knowledge that is not explicitly modeled. In the KG, teams are assigned competences and sub-competences, but projects are still missing the needed competences. Projects have some fields, such as keywords and descriptions, which is some cases contain competences.

This experiment was run 10 times with MuRE and BCEWithLogits Loss, as the model works very fast with the settings while retaining a high accuracy. MuRE yields scalar scores $y \in \Re$, with high values for y describing good fits. The model works as planned for case (1). The correct team is assigned in every run with a very high score, while all other teams are mostly predicted with a negative, or at least much worse score. Case (4) also works as intended, as all teams are predicted with low positive scores at best. The results for the other two cases vary. For the project with the competences given as keywords, the model mostly predicts the manually assigned team with a good score and sometimes with a poor score. However the intended teams are only predicted with the highest score in a few cases. The scores for the teams are mostly in the range, where an assignment would make sense, but where other teams would be a better fit. Due to the missing relation between projects and competences in the training data, this isn't necessarily bad behavior. The scores are on average highest when assigning teams to the project described as case (3), with the intended team receiving the highest score in 2 of the 10 runs.

5 Conclusion

In this paper we presented a novel approach to support SME managers in making decisions based on limited data. We have developed an automated task for assigning projects to teams with a success rate of $> 50\%$ for predicting a correct team with the highest score. Since up to 6 teams can be considered for a project, the $hits@5$ and $hits@10$ metrics are in general more important. Here we achieved results of around 90%, respectively 98% for correctly predicting a specific team. Given the small scope and the limited amount of data, the results achieved can be applied in the current use case already. However, in order to further improve the models, it is planned to add the necessary competencies to existing projects. To reduce training and query times when new projects are added to the KG, it is intended to test semi-inductive approaches based on PLMs.

Furthermore, due to the large variety of interaction models and configurations of the LP pipeline in the transductive use case, there is a possibility for further improvements in regards of the success rate. We are limited in our implementations by using models and the corresponding parts of the LP pipeline, that are already implemented by PyKEEN. Another limitation is the static nature of the initial data that is stored in Excel files. In case the company decides to restructure their teams, they will have to manually update their data, which comes down to a tiresome and imprecise process. Assisting the company in improving their data storage and handling could be a use case for future work.

KG are also suited for visualization. By converting the given data into an RDF, we supported the company in data management.

The next step is the prototype implementation of the system on the company server with further evaluation. It is also planned to look into the possibility of applying the same methodology to other SMEs and therefore validating the process.

Acknowledgments. This research and development project is funded under the funding measure "Future of Work: Regional Competence Centers for Work Research - Artificial Intelligence" in the program "Innovations for Tomorrow's Production, Services and Work" of the Federal Ministry of Education and Research (BMBF), Bonn, Germany and supervised by the Project Management Agency Karlsruhe (PTKA). Fund. No.: 02L19C500.

References

1. Ali, M., et al.: Improving inductive link prediction using hyper-relational facts. In: Hotho, A., et al. (eds.) ISWC 2021. LNCS, vol. 12922, pp. 74–92. Springer, Cham (2021). https://doi.org/10.1007/978-3-030-88361-4_5
2. Ali, M., et al.: Bringing light into the dark: a large-scale evaluation of knowledge graph embedding models under a unified framework. CoRR **abs/2006.13365** (2020). https://arxiv.org/abs/2006.13365
3. Ali, M., et al.: PyKEEN 1.0: a python library for training and evaluating knowledge graph embeddings. J. Mach. Learn. Res. **22**(82), 1–6 (2021). http://jmlr.org/papers/v22/20-825.html
4. Balažević, I., Allen, C., Hospedales, T.: Multi-relational Poincaré Graph Embeddings. Curran Associates Inc., Red Hook (2019)
5. Balazevic, I., Allen, C., Hospedales, T.M.: Tucker: tensor factorization for knowledge graph completion. In: Conference on Empirical Methods in Natural Language Processing (2019). https://api.semanticscholar.org/CorpusID:59316623
6. Bordes, A., Usunier, N., Garcia-Duran, A., Weston, J., Yakhnenko, O.: Translating embeddings for modeling multi-relational data. In: Burges, C., Bottou, L., Welling, M., Ghahramani, Z., Weinberger, K. (eds.) Advances in Neural Information Processing Systems. vol. 26. Curran Associates, Inc. (2013). https://proceedings.neurips.cc/paper_files/paper/2013/file/1cecc7a77928ca8133fa24680a88d2f9-Paper.pdf
7. Dai, Y., Wang, S., Xiong, N.N., Guo, W.: A survey on knowledge graph embedding: Approaches, applications and benchmarks. Electronics **9**(5), 750 (2020)
8. Dettmers, T., Minervini, P., Stenetorp, P., Riedel, S.: Convolutional 2D knowledge graph embeddings. In: Proceedings of the Thirty-Second AAAI Conference on Artificial Intelligence and Thirtieth Innovative Applications of Artificial Intelligence Conference and Eighth AAAI Symposium on Educational Advances in Artificial Intelligence. AAAI'18/IAAI'18/EAAI'18. AAAI Press (2018)
9. Dong, X.L., et al.: Knowledge vault: a web-scale approach to probabilistic knowledge fusion. In: The 20th ACM SIGKDD International Conference on Knowledge Discovery and Data Mining, KDD 2014, New York, NY, USA - 24 August–27, 2014, pp. 601–610 (2014). http://www.cs.cmu.edu/~nlao/publication/2014.kdd.pdf
10. Golbeck, J.: Analyzing the Social Web. Newnes (2013)
11. Hein-Pensel, F., et al.: Maturity assessment for industry 5.0: a review of existing maturity models. J. Manuf. Syst. **66**, 200–210 (2023). https://doi.org/10.1016/j.jmsy.2022.12.009, https://www.sciencedirect.com/science/article/pii/S027861252200228X
12. Khan, M., Haleem, A., Javaid, M.: Changes and improvements in industry 5.0: a strategic approach to overcome the challenges of industry 4.0. Green Technol. Sustain. **1**(2), 100020 (2023). https://doi.org/10.1016/j.grets.2023.100020, https://www.sciencedirect.com/science/article/pii/S2949736123000131
13. Li, D., et al.: Multi-task pre-training language model for semantic network completion **22**(11) (2023). https://doi.org/10.1145/3627704, https://doi.org/10.1145/3627704
14. Li, Y., et al.: Learning joint relational co-evolution in spatial-temporal knowledge graph for SMEs supply chain prediction. In: Proceedings of the 29th ACM SIGKDD Conference on Knowledge Discovery and Data Mining, pp. 4426–4436. KDD 2023. Association for Computing Machinery, New York (2023). https://doi.org/10.1145/3580305.3599855

15. Rojko, A.: Industry 4.0 concept: Background and overview. Int. J. Interact. Mob. Technol. (iJIM) **11**(5), 77-90 (2017). https://doi.org/10.3991/ijim.v11i5.7072, https://online-journals.org/index.php/i-jim/article/view/7072
16. Rossi, A., Barbosa, D., Firmani, D., Matinata, A., Merialdo, P.: Knowledge graph embedding for link prediction: a comparative analysis. ACM Trans. Knowl. Disc. Data (TKDD) **15**(2), 1–49 (2021)
17. Sun, Z., Deng, Z., Nie, J., Tang, J.: Rotate: knowledge graph embedding by relational rotation in complex space. CoRR **abs/1902.10197** (2019). http://arxiv.org/abs/1902.10197
18. Tamašauskaitė, G., Groth, P.: Defining a knowledge graph development process through a systematic review. ACM Trans. Softw. Eng. Methodol. **32**(1), 1–40 (2023)
19. Trouillon, T., Welbl, J., Riedel, S., Gaussier, E., Bouchard, G.: Complex embeddings for simple link prediction. In: Proceedings of the 33rd International Conference on International Conference on Machine Learning, vol. 48, pp. 2071–2080. ICML 2016. JMLR.org (2016)
20. Vaswani, A., et al.: Attention is all you need. CoRR **abs/1706.03762** (2017). http://arxiv.org/abs/1706.03762
21. Wang, Q., Mao, Z., Wang, B., Guo, L.: Knowledge graph embedding: A survey of approaches and applications. IEEE Trans. Knowl. Data Eng. **29**(12), 2724–2743 (2017)
22. Wang, S., et al.: Graph learning based recommender systems: a review. arXiv preprint arXiv:2105.06339 (2021)
23. Wang, Z., Zhang, J., Feng, J., Chen, Z.: Knowledge graph embedding by translating on hyperplanes. In: Proceedings of the AAAI Conference on Artificial Intelligence, vol. 28, no. 1 (2014). https://doi.org/10.1609/aaai.v28i1.8870, https://ojs.aaai.org/index.php/AAAI/article/view/8870
24. Yang, H., Lin, Z., Zhang, M.: Rethinking knowledge graph evaluation under the open-world assumption. ArXiv **abs/2209.08858** (2022). https://api.semanticscholar.org/CorpusID:252367474

Author Index

T. Villmann et al. (Eds.): WSOM+2024, LNNS 1087, pp. 227–228, 2024.
https://doi.org/10.1007/978-3-031-67159-3